Civil PE Practice Examination

Sixth Edition

Michael R. Lindeburg, PE

The Power to Pass®
www.ppi2pass.com

Professional Publications, Inc. • Belmont, California

Benefit by Registering This Book with PPI

- Get book updates and corrections.
- Hear the latest exam news.
- Obtain exclusive exam tips and strategies.
- Receive special discounts.

Register your book at **ppi2pass.com/register**.

Report Errors and View Corrections for This Book

PPI is grateful to every reader who notifies us of a possible error. Your feedback allows us to improve the quality and accuracy of our products. You can report errata and view corrections at **ppi2pass.com/errata**.

CIVIL PE PRACTICE EXAMINATION

Sixth Edition

Current printing of this edition: 1

Printing History

date	edition number	printing number	update
Jul 2012	4	2	Minor changes.
Jul 2014	5	1	New edition. Copyright update. Code updates.
Nov 2015	6	1	New edition. Copyright update. Code updates. Content changes.

Printed in the United States of America.

PPI
1250 Fifth Avenue
Belmont, CA 94002
(650) 593-9119
ppi2pass.com

ISBN: 978-1-59126-513-9

Library of Congress Control Number: 2015950983

F E D C B A

Table of Contents

Preface

By this time, if you are at a stage in your preparation where you are picking up the *Civil PE Practice Examination* and reading this preface (most people don't), then you probably have already read the prefaces in the *Civil Engineering Reference Manual* and the *Practice Problems for the Civil Engineering PE Exam*. In that case, you probably already know what changes have been incorporated into this book, since they parallel the changes to the other two books. Also, you will have read the clever and witty things that I wrote in those prefaces, so I won't be able to reuse them here.

This book is written in the multiple-choice exam format instituted by NCEES. It covers all the same topic areas that appear on the exam, as provided by NCEES. This sixth edition of the *Civil PE Practice Examination* reflects the most current NCEES exam specifications. None of the problems in this publication are actual exam problems, however. The problems in this book have come out of my head and the heads of my colleagues.

Engineering economics (a.k.a. engineering economy) is a subject that continues to appear on the exam here and there, once in a while, in one form or another. Now, with estimating and bidding being part of the construction engineering topic, engineering economics seems to have experienced a rebirth of sorts. Although the subject has been eliminated from almost all of NCEES' PE exam outlines, NCEES allows that "some problems will contain aspects of engineering economics." So, although the days are gone when an examinee had to exhibit the knowledge of a tax accountant, it is still open season on such problems.

As a final commentary on the subjects covered in this edition, I admit to being just a little bit concerned with the construction topics that lend themselves to having questions written about them. Issues of concern include the relative simplicity of estimating and quantity take-off problems, the look-up-and-search-for-the-answer aspect of OSHA safety questions, and the almost unlimited number of nonquantitative contracting, construction, and management questions that can be framed as civil engineering questions simply by including the word "engineering," "project," "contract," or "design." Oh, well. I guess all of the questions on the exam can't be based on hard-core engineering fundamentals.

In competitive soccer training, we do a drill that is basically an endless series of sprints and rests. The players call this drill "choke and puke." Tongue-in-cheek, I have given this practice exam the same name.

You are now reading the choke-and-puke edition of the *Civil PE Practice Examination*. This publication mimics the actual civil engineering PE licensing exam as closely in format, depth, and variety of questions as any exam ever published by PPI. Therein is the difficulty. This is a very difficult practice exam. It will make you want to choke and puke. So before you get discouraged from eight hours of working this practice exam, I wanted to clarify some items.

- Although you only get six minutes per problem on the real exam, you'll need more than six minutes for many of those problems. The same is true with this sample exam. Some of the problems in this practice exam are, to be honest, 12- to 15-minute problems.

- However, there are also a number of 30-second, non-quantitative questions. They are difficult, but if you know the subject material, they'll only take you the time needed to read through them. If you don't know the subject, well... choke and puke... you'll be furiously leafing through some of the references you have brought with you.

- The breadth of actual exam topics is intimidating. The knowledge you'll need to complete this practice exam will take you far afield from what is in any single book in your library. Even the fabled *Civil Engineering Reference Manual* isn't going to make up for a lack of experience in the afternoon portion of the exam. Nothing can.

- Luckily, you don't need to score 100%, 90%, or probably even 80% on the exam to pass. Take solace in knowing that you can get a bunch of the questions wrong. Nobody outside of exam administration is supposed to know what the required raw passing score is (and it's probably lower than you would think), but it's just good to know that the score is attainable by normal geeky people—you don't need to be an idiot savant.

I don't think one engineer in a hundred is going to be concerned with my editorial decisions. (So, if you are not, you can turn the page and start reading the Acknowledgments.) However, I made them (the editorial decisions), and I want you to know about them so that you don't complain when you see the real exam is different.

In writing this edition, I made several style and formatting decisions with which you might not agree. I used the same variable symbols as in the *Civil Engineering*

Reference Manual, which was modeled after the original source documents (codes and standards). The exam might use Q for flow quantity while this book uses $\dot{V}$. The exam might use w for specific weight, but this book, the *Civil Engineering Reference Manual,* and *Practice Problems for the Civil Engineering PE Exam* use γ. That shouldn't bother you.

Another decision involves the number of significant digits used in the solutions. The rules for significant digits are well known: Answers cannot be any more precise than the most imprecise parameter. Field practice often uses even fewer significant digits than would be justified by the given data, in recognition of the many unknowns and assumptions incorporated into the solution. But exam review is not field work. For this publication, I assumed you would be using a calculator and would want to compare this book's solutions with the digits appearing in your calculator. I also assumed that you would retain intermediate answers in your calculator's memory, rather than enter new numbers that you already rounded. Because of these assumptions, most of the results are printed with more significant digits than can be justified. This is intentional: you'll be able to see if you're doing it the right way. Don't be critical about the number of insignificant digits. They are there for your benefit.

Still another decision had to do with the units used in the solutions. I used common field units wherever possible, but not everywhere. As much as possible, I have tried to be consistent within this publication. Therefore, I have not used "kips" in one place and "k" in another to mean the same thing: "1000 pounds." Still, one question might be solved using "ksi," while another is solved with "psi." It all depends on the context.

In keeping with a personal preference, I was strict in differentiating between weight and mass. Thus, the units "lbf" and "lbm" are differentiated throughout, although "lb" is good enough for most civil engineering work. Similarly, I differentiated mass density, ρ, from specific weight, γ, even though the good-old "pcf" is comfortably indistinct in its meaning.

When writing and solving any metric questions in this book, I used only standard SI units. For example, you won't find any "kilogram-force" units in this book, even though they might still be in use in some countries. The SI system uses the kilogram as a unit of mass, never as weight or force.

All of the problems in this practice examination are independent. NCEES makes an attempt to "decouple" problems from previous results. For example, if question 1 asks, "What is the coefficient of active earth pressure?", then question 2 will often be phrased, "Assuming the coefficient of active earth pressure is 0.30, what is the active force on the wall?" In this manner, you still have a chance on question 2 even if you get question 1 wrong. That's good.

And bad. In some problem sets, this decoupling results in your having to repeat previous steps with the new set of given data for each subsequent problem. Although some examinees have complained about the additional workload and not being able to use the interim results in their calculator stack, many others are quite comfortable with this.

Finally, just as the *Civil Engineering Reference Manual* exceeds the scope of the actual licensing exam, this practice exam does so as well. It's about 10% more difficult than the actual exam. It's still representative and realistic, just a tiny bit more difficult.

This is a practice exam. You should use it for practice, not for prediction. You should not use this examination to predict the topics on the actual exam you should study or (heaven forbid) to predict your own performance on the actual exam. Every exam is different. Every examinee is different. Everyone has different strengths and weaknesses. Everyone has a different knowledge base and a different working speed. The passing rate for the PE exam has never been consistent. So, with each examinee being different, and with each PE exam being different, it is unlikely that I would ever be able to perfectly match the complexity and level of difficulty you experience, anyway.

As in all of my publications, I invite your comments. If you disagree with a solution, or if you think there is a better way to do something, please let me know (at **ppi2pass.com/errata**) so that I can share your comments with everyone else.

So, it's time for you to dig in and practice choking and puking your way to your engineering license. Work hard now. Sail through the exam.

Michael R. Lindeburg, PE

Acknowledgments

I owe a huge debt to a number of people who kept this publication on track with the current exam, the latest codes and standards, and modern practice. For the sixth edition, I depended upon subject matter experts from industry and academia to review existing material: Akash Rao, PE, LEED BD+C, Keith Warwick, PE, and Ralph Arcena. Akash Rao revised the structural chapters in accordance with the exam-adopted structural codes, and Keith Warwick revised the construction chapters in accordance with the exam-adopted construction codes. Ralph Arcena calculation checked problems throughout this edition.

The staff that worked on this edition is the best trained and most proficient that PPI has ever had. Managed by Sarah Hubbard, director of product development and implementation, Sam Webster, editorial manager, Serena Cavanaugh, acquisitions editor, and Cathy Schrott, production services manager, the following top-drawer individuals have my thanks for bringing this book to life: Jennifer Lindeburg King, associate editor-in-chief; Nicole Evans, associate project manager; Thomas Bliss, Sierra Cirimelli-Low, Tyler Hayes, Tracy Katz, Scott Marley, Ellen Nordman, and Ian A. Walker, copy editors; and Tom Bergstrom, production associate and technical illustrator.

Michael R. Lindeburg, PE

Introduction

ABOUT THE PE EXAM

Civil PE Practice Examination provides the opportunity to practice taking an eight-hour test similar in content and format to the Principles and Practice of Engineering (PE) examination in civil engineering. The civil PE examination is an eight-hour exam divided into a morning session and an afternoon session. The morning session is known as the "breadth" exam, and the afternoon session is known as the "depth" exam. This book contains a sample breadth module and five sample depth modules—one for each subdiscipline tested by NCEES.

In the four-hour morning session, the examinee is asked to solve 40 problems from eight knowledge areas: project planning, means and methods, soil mechanics, structural mechanics, hydraulics and hydrology, geometrics, materials, and site development. Morning session problems are general in nature and wide-ranging in scope.

The afternoon session is also four hours. NCEES allows the examinee to select one of five depth modules (construction; geotechnical; structural; transportation; or water resources and environmental) when registering for the exam. At the exam, only the examination booklet of the depth module selected when registering for the exam will be provided. Switching modules is not possible. The answer sheet will be scored based on the module selected during registration. Each depth module is made up of 40 problems. Afternoon session problems require more specialized knowledge than those in the morning session.

All problems, from both the morning and afternoon sessions, are multiple choice. They include a problem statement with all required defining information, followed by four logical choices. Only one of the four options is correct. The problems are completely independent of each other, so an incorrect choice on one problem will not carry over to subsequent problems.

Topics and the approximate distribution of problems on the morning session of the exam are as follows.

- **Project Planning (4 problems)**

  Quantity take-off methods; cost estimating; project schedules; activity identification and sequencing

- **Means and Methods (3 problems)**

  Construction loads; construction methods; temporary structures and facilities

- **Soil Mechanics (6 problems)**

  Lateral earth pressure; soil consolidation; effective and total stresses; bearing capacity; foundation settlement; slope stability

- **Structural Mechanics (6 problems)**

  Dead and live loads; trusses; bending moments and stresses; shear forces and stresses; axial forces and stresses; combined stresses; deflection; beams; columns; slabs; footings; retaining walls

- **Hydraulics and Hydrology (7 problems)**

  Open-channel flow; stormwater collection and drainage; storm characteristics; runoff analysis; detention/retention ponds; pressure conduit; energy and/or continuity equation

- **Geometrics (3 problems)**

  Basic circular curve elements; basic vertical curve elements; traffic volume

- **Materials (6 problems)**

  Soil classification and boring log interpretation; soil properties; concrete; structural steel; material test methods and specification conformance; compaction

- **Site Development (5 problems)**

  Excavation and embankment; construction site layout and control; temporary and permanent soil erosion and sediment control; impact of construction on adjacent facilities; safety

Topics and the approximate distribution of problems in the five depth modules of the civil PE exam afternoon session are as follows.

Construction Module

- Earthwork construction and layout (6 problems)
- Estimating quantities and costs (6 problems)
- Construction operations and methods (7 problems)
- Scheduling (5 problems)
- Material quality control and production (6 problems)
- Temporary structures (7 problems)
- Health and safety (3 problems)

Geotechnical Module

- Site characterization (5 problems)

- Soil mechanics, laboratory testing, and analysis (5 problems)

- Field materials testing, methods, and safety (3 problems)

- Earthquake engineering and dynamic loads (2 problems)

- Earth structures (4 problems)

- Groundwater and seepage (3 problems)

- Problematic soil and rock conditions (3 problems)

- Earth retaining structures (5 problems)

- Shallow foundations (5 problems)

- Deep foundations (5 problems)

Structural Module

- Analysis of structures (14 problems)

- Design and details of structures (20 problems)

- Codes and construction (6 problems)

Transportation Module

- Traffic engineering (11 problems)

- Horizontal design (4 problems)

- Vertical design (4 problems)

- Intersection geometry (4 problems)

- Roadside and cross-section design (4 problems)

- Signal design (3 problems)

- Traffic control design (3 problems)

- Geotechnical and pavement (4 problems)

- Drainage (2 problems)

- Alternative analysis (1 problem)

Water Resources and Environmental Module

- Project planning (4 problems)

- Means and methods (3 problems)

- Soil mechanics (6 problems)

- Structural mechanics (6 problems)

- Hydraulics and hydrology (7 problems)

- Geometrics (3 problems)

- Materials (6 problems)

- Site development (5 problems)

According to NCEES, exam questions related to codes and standards will be based on either (1) an interpretation of a code or standard that is presented in the exam booklet or (2) a code or standard that a committee of licensed engineers feels minimally competent engineers should know. Code information required to solve questions will be consistent with the edition of the code adopted by NCEES for the exam.

For further information and tips on how to prepare for the civil PE exam, consult the *Civil Engineering Reference Manual* or PPI's website at **ppi2pass.com/cefaq**.

HOW TO USE THIS BOOK

Warning Advisories: (1) The problems in this practice examination evolved out of practice exams for three-month review courses, and were intended to be challenging and thought-provoking even for the participants of the review courses. If you have not spent hundreds of hours over the previous months becoming intimately familiar with the *Civil Engineering Reference Manual* (CERM) and working through the problems in *Practice Problems for the Civil Engineering PE Exam* (CEPP), this practice exam will be more than challenging. It will be decidedly frustrating for you. Not only will many of the subjects be unfamiliar, but you won't be able to solve many of the problems in six minutes. (2) This practice exam was designed to require an average time of approximately six minutes per problem. The intention is not that each problem can be solved in six minutes. Like all exams, some problems are basic and some are complex. Some problems are intentionally short, with the accumulated remaining time intended for use on the long problems. If you are unprepared for the exam, or are not familiar with CERM, or if you have not worked through the problems in CEPP, most of the problems in this practice exam will seem unrealistically long.

It is recommended that you treat this book as an exam. Do not read the questions ahead of time, and do not look at the answers until you've finished. As you work the problems, you may use the *Civil Engineering Reference Manual*. Adequate preparation, not an extensive library, is the key to success. Check with your state's board of engineering registration for any restrictions. (The PPI website has a listing of state boards at **ppi2pass.com/stateboards**.)

Prepare for the exam, read the practice exam instructions (which simulate the ones you'll receive from your exam proctor), set a timer for four hours, and take the breadth module. After a one-hour break, turn to the depth module you have chosen for the actual exam, set the timer, and complete the simulated afternoon session. Then, check your answers.

After taking the practice exam, review your areas of weakness and then take the exam again, but since none of the problems in the book are repeated, substitute a different depth module. Check your answers, and repeat the process for each of the depth areas. Evaluate your strengths and weaknesses, and select additional texts to supplement your weak areas (e.g., *Practice Problems for the Civil Engineering PE Exam*). Check the PPI website for the latest in exam preparation materials at **ppi2pass.com**.

The problems in this book were written to emphasize the breadth of the civil engineering field. Some may seem easy and some hard. If you are unable to answer a given question, you should review that topic area.

This book assumes that the breadth module of the PE exam will be more academic and traditional in nature, and that the depth modules will require practical, non-numerical knowledge of the type that comes from experience.

The problems are generally similar to each other in difficulty, yet a few somewhat easier problems have been included to expose you to less-frequently examined topics.

The keys to success on the exam are to know the basics and to practice solving as many problems as possible. This book will assist you with both objectives.

HOW TO INTERPRET YOUR PRACTICE EXAMINATION PERFORMANCE

The average time allowed for each problem during the actual examination is six minutes. You might assume that you will need six minutes to solve each of the problems. Unfortunately (particularly if you prepare minimally), there are two factors that may disappoint or frustrate you if you are expecting to solve each problem in six minutes or less.

The first factor is your own speed. Your average speed may not coincide with the examination's average speed. If you have prepared thoroughly over 4–6 months, you might power through the examination and finish with a little time to spare. However, most examinees cannot finish the examination, even if they rush. If you are underprepared, don't have the references needed, or just work slowly, you'll end up guessing at a lot of problems during the last 5 minutes.

The second factor is the diversity of problem difficulties on the exam. While the hypothetical average examinee is allowed an average of six minutes per problem, some qualitative problems (e.g., definitions, concepts, and theories such as "What type of weir is most useful with head waters of varying depth?") require only 30 seconds to solve. When you encounter one of these, you'll be able to read it, think about it, and choose the correct answer option quickly.

Quantitative problems (i.e., those requiring calculations) take longer, some taking as much as 20 minutes of your time. A few quantitative problems will require as few as one or two calculations, such as "Water flows in a 6 in pipe at 5.0 ft/sec. What is most nearly the flow rate?" Other problems will require an "endless string" of consecutive steps, such as when you to have to (1) determine fluid density and viscosity based on the temperature, (2) convert the mass flow rate to a volumetric flow rate, (3) determine the pipe diameter from the pipe size designation (e.g., pipe schedule), (4) calculate the internal pipe area, (5) calculate the flow velocity, (6) determine the specific roughness from the conduit material, (7) calculate the relative roughness, (8) calculate the Reynolds number, (9) calculate or determine the friction factor graphically, (10) determine the equivalent length of fittings and other minor losses, (11) calculate the head loss, and finally, (12) convert the head loss to pressure drop.

The problems in this practice exam contain several of those "endless string" problems that take 20 minutes to solve. If you take the practice exam under realistic, timed conditions, you may be unable to finish it in the allotted time, or you may find that you are performing at the 45–55% level when you grade it. 50% is adequate performance if you take the practice exam under realistic, timed conditions. If you give yourself twice as much time, your performance should be 70%.

If you are frustrated, discouraged, disappointed, angry, or worried after working through this practice exam, you may take comfort in knowing that some of the problems on the actual exam will require only one or two of the steps in the "endless string" sequences. By working these harder problems, you'll be well prepared to solve those one or two step problems during the actual exam. However, some of the problems will take 10 minutes, and some will take 20 minutes, just like the problems in this practice exam.

Codes and Standards

The information that was used to write and update this book was based on the exam specifications at the time of publication. However, as with engineering practice itself, the PE exam is not always based on the most current codes or cutting-edge technology. Similarly, codes, standards, and regulations adopted by state and local agencies often lag issuance by several years. It is likely that the codes that are most current, the codes that you use in practice, and the codes that are the basis of your exam will all be different.

PPI lists on its website the dates and editions of the codes, standards, and regulations on which NCEES has announced the PE exams are based. It is your responsibility to find out which codes are relevant to your exam.

CONSTRUCTION DESIGN STANDARDS

ACI 318: *Building Code Requirements for Structural Concrete*, 2011. American Concrete Institute, Farmington Hills, MI.

ACI 347: *Guide to Formwork for Concrete*, 2004. American Concrete Institute, Farmington Hills, MI. (In ACI SP-4, Seventh ed. App.)

ACI SP-4: *Formwork for Concrete*, Seventh ed., 2005. American Concrete Institute, Farmington Hills, MI.

AISC: *Steel Construction Manual*, Fourteenth ed., 2011. American Institute of Steel Construction, Inc., Chicago, IL.

ASCE 37: *Design Loads on Structures During Construction*, 2002. American Society of Civil Engineers, Reston, VA.

CMWB: *Standard Practice for Bracing Masonry Walls Under Construction*, 2012. Council for Masonry Wall Bracing, Mason Contractors Association of America, Lombard, IL.

MUTCD-Pt 6: *Manual on Uniform Traffic Control Devices*—Part 6, Temporary Traffic Control, 2009. U.S. Department of Transportation, Federal Highway Administration, Washington, DC.

NDS: *National Design Specification for Wood Construction ASD/LRFD*, 2012 ed. American Wood Council, Washington, DC.

OSHA 1926: *Occupational Safety and Health Regulations for the Construction Industry* (U.S. Federal version). U.S. Department of Labor, Washington, DC.

GEOTECHNICAL DESIGN STANDARDS

ASCE/SEI7: *Minimum Design Loads for Buildings and Other Structures*, 2010. American Society of Civil Engineers, Reston, VA.

OSHA 1926: *Occupational Safety and Health Regulations for the Construction Industry* (U.S. Federal version). U.S. Department of Labor, Washington, DC.

STRUCTURAL DESIGN STANDARDS

AASHTO *LRFD: AASHTO LRFD Bridge Design Specifications*, Sixth ed., 2012. American Association of State Highway and Transportation Officials, Washington, DC.

ACI 318[1]: *Building Code Requirements for Structural Concrete*, 2011. American Concrete Institute, Farmington Hills, MI.

ACI 530/530.1[2]: *Building Code Requirements and Specification for Masonry Structures* (and companion commentaries), 2011. The Masonry Society, Boulder, CO; American Concrete Institute, Detroit, MI; and Structural Engineering Institute of the American Society of Civil Engineers, Reston, VA.

AISC: *Steel Construction Manual*, Fourteenth ed., 2011. American Institute of Steel Construction, Inc., Chicago, IL.

ASCE/SEI7: *Minimum Design Loads for Buildings and Other Structures*, 2010. American Society of Civil Engineers, Reston, VA.

AWS D1.1/D1.1M[3]: *Structural Welding Code—Steel*, Twenty-second ed., 2010. American Welding Society, Miami, FL.

AWS D1.2/D1.2M[4]: *Structural Welding Code—Aluminum*, Sixth ed., 2014. American Welding Society, Miami, FL.

---

[1]ACI 318 App. C does not apply to the Civil PE structural depth exam.

[2]Only the Allowable Stress Design (ASD) method may be used on the exam, except that ACI 530 Sec. 3.3.5 (strength design) may be used for walls with out-of-plane loads.

[3]AWS D1.1, AWS D1.2, and AWS D1.4 are listed in the Codes, Standards, and Documents subsection of NCEES's Civil PE structural depth exam specifications.

[4]See Ftn. 3.

AWS D1.4/D1.4M[5]: *Structural Welding Code—Reinforcing Steel*, Seventh ed., 2011. American Welding Society, Miami, FL.

IBC: *2012 International Building Code* (without supplements). International Code Council, Inc., Falls Church, VA.

NDS[6]: *National Design Specification for Wood Construction ASD/LRFD*, 2012 ed., and *National Design Specification Supplement, Design Values for Wood Construction*, 2012 ed. American Wood Council, Washington, DC.

OSHA 1910[7]: *Occupational Safety and Health Standards* (U.S. Federal version). U.S. Department of Labor, Washington, DC.

OSHA 1926: *Occupational Safety and Health Regulations for the Construction Industry* (U.S. Federal version) Subpart E, Personal Protective and Life Saving Equipment, 1926.95–1926.107; Subpart M, Fall Protection, 1926.500–1926.503, App. A–E; Subpart Q, Concrete and Masonry Construction, 1926.700–1926.706, with App. A; and Subpart R, Steel Erection, 1926.750–1926.761, with App. A–H. U.S. Department of Labor, Washington, DC.

PCI: *PCI Design Handbook: Precast and Prestressed Concrete*, Seventh ed., 2010. Precast/Prestressed Concrete Institute, Chicago, IL.

TRANSPORTATION DESIGN STANDARDS

AASHTO *GDPS*: *AASHTO Guide for Design of Pavement Structures* (GDPS-4-M), 1993, and 1998 supplement. American Association of State Highway and Transportation Officials, Washington, DC.

AASHTO *Green Book*: *A Policy on Geometric Design of Highways and Streets*, Sixth ed., 2011. American Association of State Highway and Transportation Officials, Washington, DC.

AASHTO: *Guide for the Planning, Design, and Operation of Pedestrian Facilities*, First ed., 2004. American Association of State Highway and Transportation Officials, Washington, DC.

HSM: *Highway Safety Manual*, First ed., 2010. American Association of State Highway and Transportation Officials, Washington, DC.

AASHTO *MEPDG*: *Mechanistic-Empirical Pavement Design Guide: A Manual of Practice*, Interim ed., 2008. American Association of State Highway and Transportation Officials, Washington, DC.

AASHTO: *Roadside Design Guide*, Fourth ed., 2011. American Association of State Highway and Transportation Officials, Washington, DC.

AI: *The Asphalt Handbook* (MS-4), Seventh ed., 2007. Asphalt Institute, Lexington, KY.

FHWA: *Hydraulic Design of Highway Culverts*, Hydraulic Design Series no. 5, Publication no. FHWA-HIF-12-026, Third ed., 2012. U.S. Department of Transportation, Federal Highway Administration, Washington, DC.

HCM: *Highway Capacity Manual*, 2010 ed. Transportation Research Board, National Research Council, Washington, DC.

MUTCD: *Manual on Uniform Traffic Control Devices*, 2009 (including Revisions 1 and 2, May 2012). U.S. Department of Transportation, Federal Highway Administration, Washington, DC.

PCA: *Design and Control of Concrete Mixtures*, Fifteenth ed., 2011. Portland Cement Association, Skokie, IL.

---

[5]See Ftn. 3.
[6]Only the ASD method may be used for wood design on the exam.
[7]Part 1910 is listed in the Codes, Standards, and Documents subsection of NCEES's Civil PE structural depth exam specifications.

Morning Session

In accordance with the rules established by your state, you may use textbooks, handbooks, bound reference materials, and any approved battery- or solar-powered, silent calculator to work this examination. However, no blank papers, writing tablets, unbound scratch paper, or loose notes are permitted. Sufficient room for scratch work is provided in the Examination Booklet.

You are not permitted to share or exchange materials with other examinees. However, the books and other resources used in this morning session may be changed prior to the afternoon session.

You will have four hours in which to work this session of the examination. Your score will be determined by the number of questions that you answer correctly. There is a total of 40 questions. All 40 questions must be worked correctly in order to receive full credit on the exam. There are no optional questions. Each question is worth 1 point. The maximum possible score for this section of the examination is 40 points.

Partial credit is not available. No credit will be given for methodology, assumptions, or work written in your Examination Booklet.

Record all of your answers on the Answer Sheet. No credit will be given for answers marked in the Examination Booklet. Mark your answers with the official examination pencil provided to you. Marks must be dark and must completely fill the bubbles. Record only one answer per question. If you mark more than one answer, you will not receive credit for the question. If you change an answer, be sure the old bubble is erased completely; incomplete erasures may be misinterpreted as answers.

If you finish early, check your work and make sure that you have followed all instructions. After checking your answers, you may turn in your Examination Booklet and Answer Sheet and leave the examination room. Once you leave, you will not be permitted to return to work or change your answers.

When permission has been given by your proctor, break the seal on the Examination Booklet. Check that all pages are present and legible. If any part of your Examination Booklet is missing, your proctor will issue you a new Booklet.

WAIT FOR PERMISSION TO BEGIN

Name: _____
      Last   First   Middle Initial

Examinee number: _____

Examination Booklet number: _____

Principles and Practice of Engineering Examination

Morning Session Practice Examination

Morning Session

1.	Ⓐ Ⓑ Ⓒ Ⓓ	11. Ⓐ Ⓑ Ⓒ Ⓓ	21. Ⓐ Ⓑ Ⓒ Ⓓ	31. Ⓐ Ⓑ Ⓒ Ⓓ
2.	Ⓐ Ⓑ Ⓒ Ⓓ	12. Ⓐ Ⓑ Ⓒ Ⓓ	22. Ⓐ Ⓑ Ⓒ Ⓓ	32. Ⓐ Ⓑ Ⓒ Ⓓ
3.	Ⓐ Ⓑ Ⓒ Ⓓ	13. Ⓐ Ⓑ Ⓒ Ⓓ	23. Ⓐ Ⓑ Ⓒ Ⓓ	33. Ⓐ Ⓑ Ⓒ Ⓓ
4.	Ⓐ Ⓑ Ⓒ Ⓓ	14. Ⓐ Ⓑ Ⓒ Ⓓ	24. Ⓐ Ⓑ Ⓒ Ⓓ	34. Ⓐ Ⓑ Ⓒ Ⓓ
5.	Ⓐ Ⓑ Ⓒ Ⓓ	15. Ⓐ Ⓑ Ⓒ Ⓓ	25. Ⓐ Ⓑ Ⓒ Ⓓ	35. Ⓐ Ⓑ Ⓒ Ⓓ
6.	Ⓐ Ⓑ Ⓒ Ⓓ	16. Ⓐ Ⓑ Ⓒ Ⓓ	26. Ⓐ Ⓑ Ⓒ Ⓓ	36. Ⓐ Ⓑ Ⓒ Ⓓ
7.	Ⓐ Ⓑ Ⓒ Ⓓ	17. Ⓐ Ⓑ Ⓒ Ⓓ	27. Ⓐ Ⓑ Ⓒ Ⓓ	37. Ⓐ Ⓑ Ⓒ Ⓓ
8.	Ⓐ Ⓑ Ⓒ Ⓓ	18. Ⓐ Ⓑ Ⓒ Ⓓ	28. Ⓐ Ⓑ Ⓒ Ⓓ	38. Ⓐ Ⓑ Ⓒ Ⓓ
9.	Ⓐ Ⓑ Ⓒ Ⓓ	19. Ⓐ Ⓑ Ⓒ Ⓓ	29. Ⓐ Ⓑ Ⓒ Ⓓ	39. Ⓐ Ⓑ Ⓒ Ⓓ
10.	Ⓐ Ⓑ Ⓒ Ⓓ	20. Ⓐ Ⓑ Ⓒ Ⓓ	30. Ⓐ Ⓑ Ⓒ Ⓓ	40. Ⓐ Ⓑ Ⓒ Ⓓ

Morning Session

1. Consider the borrow pit grid shown. Existing excavation depths (in feet) are shown adjacent to each corner.

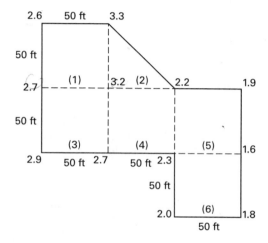

The total undercut volume of this grid is most nearly

(A) 1300 yd³

(B) 1600 yd³

(C) 1900 yd³

(D) 2100 yd³

2. A given pozzolan contains approximately 20% calcium oxide and less than 2% carbon, and has a 0.5% loss on ignition. This material is most likely to be

(A) slag

(B) class C flyash

(C) class F flyash

(D) silica fume

p. 48-2

3. The design criteria for a normalweight concrete mix are as follows.

- maximum size of coarse aggregate (CA): ³⁄₄ in

- water/cement ratio: 0.5

- slump for interior building columns: 3 in to 4 in

- specific gravity of aggregates: 2.70

- specific gravity of cement: 3.15

- dry rodded bulk specific weight of CA: 100 lbf/ft³

- fineness modulus (fm): 2.80, based on ASTM C125

- water required based on slump and CA size:

slump (in)	mix water for ¹⁄₂ in CA (lbf/yd³)	mix water for ³⁄₄ in CA (lbf/yd³)
1–2	270	250
3–4	320	300
5–6	350	320

- bulk volume of dry rodded CA based on various fineness moduli (fm):

maximum size of CA (in)	fm = 2.6 (ft³/ft³)	fm = 2.7 (ft³/ft³)	fm = 2.8 (ft³/ft³)
¹⁄₂	0.55	0.52	0.50
³⁄₄	0.65	0.62	0.60
1	0.70	0.68	0.65

For 27 ft³ of concrete mix, the absolute volumes of the water, cement, coarse aggregate (CA), and fine aggregate (FA) are most nearly

(A) 4.25 ft³ water, 3.05 ft³ cement, 10.46 ft³ CA, 9.24 ft³ FA

(B) 4.81 ft³ water, 1.52 ft³ cement, 10.46 ft³ CA, 10.21 ft³ FA

(C) 4.25 ft³ water, 1.52 ft³ cement, 11.27 ft³ CA, 9.96 ft³ FA

(D) 4.81 ft³ water, 3.05 ft³ cement, 9.62 ft³ CA, 9.52 ft³ FA

4. A small construction firm receives a $100,000 check for recently sold equipment. The owner decides to invest the money for 10 years in an account with an 8% interest rate. In addition, $20,000 is invested every year at the same interest rate. In 10 years, the total amount of money in the account will be most nearly

(A) $259,000

(B) $318,000

(C) $420,000

(D) $506,000

5. Float is defined as the

(A) continuation of an activity with the longest duration

(B) time extension of an activity with the shortest duration

(C) time difference between late finish and late start

(D) the amount of time an activity can be delayed without delaying any succeeding activities

6. Consider the following precedence table and critical path network.

activity	duration (days)
A	3
B	4
C	5
D	7
E	8
F	9
G	6

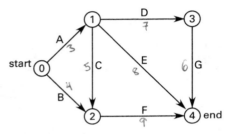

How many days are needed to complete this project?

(A) 16 days

(B) 17 days

(C) 22 days

(D) 25 days

7. Consider the soldier beam shoring system shown. It is anchored to a soft limestone material through a circular 12 ft long, 6 in diameter diagonal shaft. The bond strength between the anchor shaft and the limestone is 120 psi.

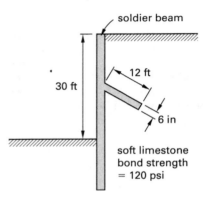

The unfactored anchor capacity of the tieback system is most nearly

(A) 75 kips

(B) 150 kips

(C) 320 kips

(D) 650 kips

8. A large backhoe bucket has a capacity of 2.0 yd^3 and a cycle time of 40 sec. The production volume in a 10 hour day with a one-hour lunch break and two 15 minute breaks is most nearly

(A) 640 yd^3

(B) 1300 yd^3

(C) 1500 yd^3

(D) 1800 yd^3

9. A gravity retaining wall is holding soil backfill with the properties shown. The total active force per unit length of the wall is most nearly

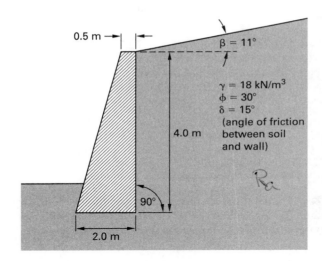

(A) 40 kN/m

(B) 50 kN/m

(C) 70 kN/m

(D) 80 kN/m

10. A sand fill is spread on top of organic silt as shown. Assume that the fill is infinite in extent. At the end of the consolidation process, the increase in vertical effective stress at point A due to the placement of the sand fill is most nearly

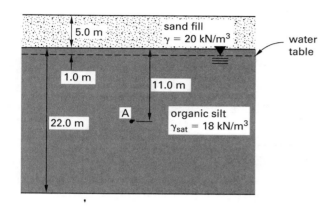

(A) 0 kN/m^2

(B) 60 kN/m^2

(C) 100 kN/m^2

(D) 120 kN/m^2

11. An impermeable dam impounds water over a soil with properties as shown. The width of the dam across the water flow is 300 m. The quantity of seepage under the dam is most nearly

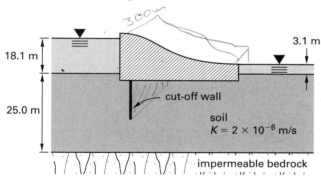

(A) 1.5×10^{-6} m^3/s

(B) 2.5×10^{-6} m^3/s

(C) 1.5×10^{-3} m^3/s

(D) 2.5×10^{-3} m^3/s

12. A square foundation is to support a column load of 800 kN. The soil beneath the footing is generally homogeneous. If the foundation bearing pressure from this load is reduced from 400 kPa to 100 kPa by increasing the foundation area (the column load remaining constant), the change in stress at a depth of 3 m below the foundation center will be most nearly

(A) a decrease in stress of 20 kPa

(B) a decrease in stress of 10 kPa

(C) an increase in stress of 10 kPa

(D) an increase in stress of 20 kPa

13. A retaining wall is shown. For the given conditions, the factor of safety against overturning is most nearly

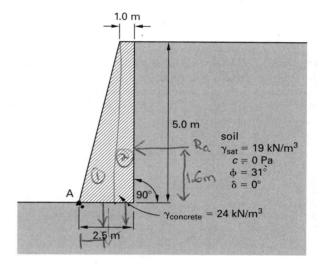

(A) 1.3

(B) 2.3

(C) 2.5

(D) 5.4

14. During excavation in saturated soil for a building foundation, a slope with the soil properties given was created. Assuming toe-slope failure, the cohesive factor of safety for the stability of this slope is most nearly

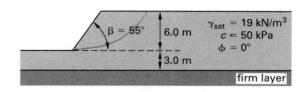

(A) 1.0

(B) 1.5

(C) 2.5

(D) 5.0

15. The results from a series of direct shear tests on a sandy soil are shown in the following table.

test number	normal stress (kPa) σ	shear stress (kPa) τ
1	50	36
2	150	105
3	250	175

The principal stresses on the failure plane for test 2 are most nearly

(A) $\sigma_1 = 100$ kPa and $\sigma_3 = 250$ kPa

(B) $\sigma_1 = 100$ kPa and $\sigma_3 = 350$ kPa

(C) $\sigma_1 = 150$ kPa and $\sigma_3 = 390$ kPa

(D) $\sigma_1 = 150$ kPa and $\sigma_3 = 490$ kPa

16. Sieve and hydrometer testing shows that a soil has the following grain size distribution. The material passing through a no. 40 sieve has a liquid limit of 34 and a plasticity index of 13. The AASHTO classification for this soil is

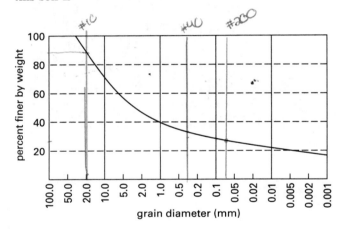

(A) A-2-6 (0)

(B) A-2-6 (1)

(C) A-2-7 (0)

(D) A-2-7 (1)

0.425mm

200:% finer 28%

#10: 90%

#40. 37%

17. A simply supported girder spans 80 ft and is subjected to a set of three moving wheel loads with magnitude and spacing as shown. What is most nearly the absolute maximum bending moment caused by the moving loads?

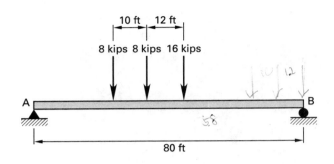

(A) 510 ft-kips

(B) 590 ft-kips

(C) 650 ft-kips

(D) 740 ft-kips

18. What is most nearly the compressive force in member CD in the truss shown, where tension is positive and compression is negative?

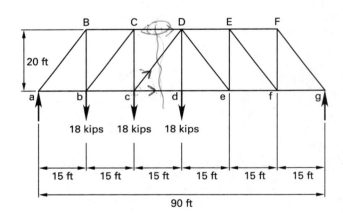

(A) −90 kips

(B) −60 kips

(C) −40 kips

(D) −20 kips

19. A solid masonry column has a cross section measuring 32 in × 32 in. The column is subjected to axial compression force of 110 kips, which includes the weight of the column, with an eccentricity of 5 in about one axis. What is most nearly the maximum compression stress caused by this loading?

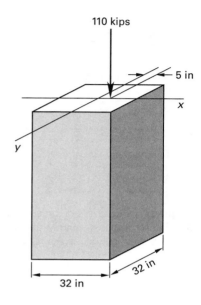

110 kips

5 in

x

y

32 in

32 in

(A) 0.2 kips/in²
(B) 1.3 kips/in²
(C) 2.4 kips/in²
(D) 3.0 kips/in²

20. A singly reinforced concrete beam has the cross section shown. The concrete is normalweight with a specified compressive strength of 4000 psi and is reinforced with four no. 10, grade 60 rebars. Normal vertical stirrups are used for shear reinforcements. The design moment strength of the section is most nearly

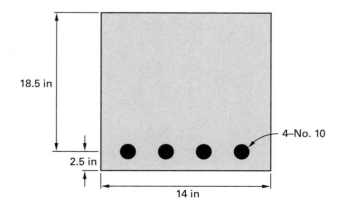

18.5 in

2.5 in

14 in

4–No. 10

(A) 330 ft-kips
(B) 350 ft-kips
(C) 390 ft-kips
(D) 420 ft-kips

21. A 14 in × 14 in reinforced concrete column bears on a square spread footing that is 8.5 ft × 8.5 ft in plan, has an overall thickness of 20 in, and is reinforced with no. 8 rebars in each direction. The footing is constructed of 3000 psi normalweight concrete in accordance with the ACI 318 specification. The maximum concentric design axial force that could be supported by the footing based on its punching shear resistance is most nearly

(A) 350 kips
(B) 400 kips
(C) 450 kips
(D) 500 kips

22. An eccentrically loaded connection is made using high-strength bolts of the same size in the arrangement shown. Based on linear elastic theory, the maximum shear that occurs in the fastener group caused by the applied force is most nearly

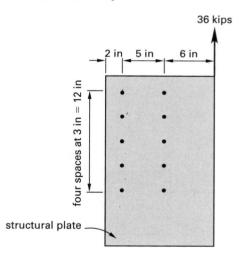

36 kips

2 in 5 in 6 in

four spaces at 3 in = 12 in

structural plate

(A) 4 kips
(B) 8 kips
(C) 10 kips
(D) 20 kips

23. A flexible plywood diaphragm spans shear walls located on lines A, B, and C, and is subjected to a lateral wind load of 320 lbf/ft acting in the direction shown. The maximum chord force created in the diaphragm is most nearly

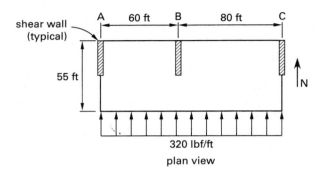

320 lbf/ft

plan view

(A) 4200 lbf

(B) 4700 lbf

(C) 9600 lbf

(D) 12,000 lbf

24. Loads on a highly restrained connection result in a state of stress having equal tensile stresses on three orthogonal faces. The connection is made by welding a ductile structural steel using an appropriate electrode. Given that loads increase until failure initiates at the stressed point, the resulting failure would be best described as

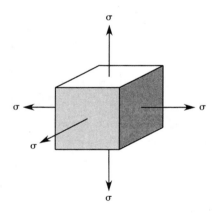

(A) ductile

(B) shear slip

(C) cleavage

(D) large deformation

25. Two tangents are connected by a circular curve as shown. The circular curve represents the centerline of a two-lane rural highway that passes through level surroundings. The highway has 12 ft lanes with 6 ft shoulders. The curve degree of curvature is 2°, and the design speed is 40 mph. What is most nearly the bearing of the radius line meeting the curve at PT?

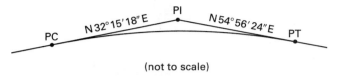

(not to scale)

(A) N 35° 3′ 36″ E

(B) N 54° 56′ 24″ E

(C) S 34° 2′ 16″ E

(D) S 35° 3′ 36″ E

26. An unsignalized T-intersection's three legs are each two way. What is most nearly the total number of conflict points, including crossing, merging, and diverging?

(A) 3

(B) 6

(C) 9

(D) 12

27. A four-lane freeway runs through rural areas. Each lane is 11 ft wide. A recent traffic study for a particular portion of the daily commute period shows the directional weekday volume is 2400 vph in one direction. An average of 750 vehicles passes by during the busiest 15 min. What is most nearly the peak hour factor (PHF)?

(A) 0.31

(B) 0.80

(C) 0.94

(D) 1.0

28. The centerline of a two-lane rural highway includes a circular curve as shown. The highway has 12 ft lanes with 6 ft shoulders. The curve degree of curvature is 2°, and the design speed is 40 mph. The curve is in the vicinity of an established Civil War cemetery. The point on the curve closest to the corner of the cemetery has coordinates 424,180.59 N ft and 268,549.70 E ft. What is most nearly the distance from the curve to the corner of the cemetery?

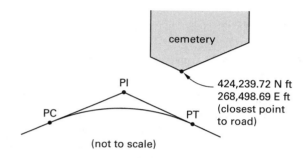

424,239.72 N ft
268,498.69 E ft
(closest point to road)

(not to scale)

(A) 77.5 ft

(B) 77.8 ft

(C) 78.1 ft

(D) 78.4 ft

29. Which one of the following statements is true?

(A) ADT is the average of 24 hour traffic counts collected every day in the year.

(B) Fixed traffic delay on roadways is caused by traffic side friction.

(C) Space mean speed is always less than or equal to mean speed.

(D) Local streets provide more access than mobility, and they carry more than 80% of travel volume nationwide.

30. A car is traveling on a two-lane rural road at 45 mph. The road grade is 5% downhill. A deer appears in front of the car and starts to cross the road. What is most nearly the distance the car needs in order to stop in time to avoid hitting the deer?

(A) 150 ft

(B) 170 ft

(C) 390 ft

(D) 420 ft

31. In the following table, ADT data for traffic movements between four locations are given. Points A, B, C, and D represent the locations along various straight highway sections, as shown in the illustration. AB represents the number of daily trips from location A to location B. BA represents the number of daily trips from B to A. Other combinations of A, B, C, and D are interpreted similarly. An interchange is proposed to accommodate the traffic to and from all locations. What is the most suitable type of interchange?

turning movement	volume (vpd)
AB	18,500
AC	25
AD	15
BA	17,000
BD	30
BC	10
CD	90
CA	15
CB	20
DC	120
DB	25
DA	20

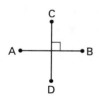

(A) cloverleaf, full

(B) cloverleaf, partial

(C) diamond

(D) directional

32. The total cut area and total fill area for two stations (1 and 2) along a roadway are as follows.

station	total cut area (ft²)	total fill area (ft²)
1	9	24
2	15	32

Use the average end area method for earthwork computations using 100 ft stations. What is most nearly the net earthwork volume?

(A) cut: 45 yd³

(B) fill: 60 yd³

(C) fill: 100 yd³

(D) cut: 150 yd³

33. A complete-mix activated sludge process is used to treat 8 MGD of brewery wastes with a COD of 1800 mg/L. The nonbiodegradable fraction is 110 mg/L COD. The biochemical reaction is pseudo-first order. The substrate utilization rate constant based on mixed liquor volatile suspended solids (MLVSS) is 0.6 L/g·h at 20°C. The design mixed liquor suspended solids (MLSS) is 2500 mg/L, and the solids are 75% volatile. Activated sludge is returned directly to the reactor for a design effluent COD of 200 mg/L. The reactor volume is most nearly

(A) 200,000 ft^3

(B) 400,000 ft^3

(C) 500,000 ft^3

(D) 700,000 ft^3

34. A culvert system is being designed to pass under a major highway. The culvert system must be able to protect the highway from runoff from a 1 in storm. The following information has been derived from a storm that produced runoff over a 2 hr period.

drainage area	43 mi^2
flood hydrograph peak discharge	9300 ft^3/sec
flood hydrograph volume	3260 ac-ft

What is most nearly the 2 hr unit hydrograph peak discharge?

(A) 2300 ft^3/in-sec

(B) 3300 ft^3/in-sec

(C) 4800 ft^3/in-sec

(D) 6500 ft^3/in-sec

35. A spray system is designed as part of an industrial process. A 0.8 in diameter nozzle connected to a 3.15 in diameter pipe will provide a spray velocity of 98 ft/sec. The coefficients of velocity and contraction are 0.95 and 0.80, respectively. The water temperature is 160°F. The pressure required at the entrance to the nozzle is most nearly

(A) 46 psi

(B) 54 psi

(C) 70 psi

(D) 77 psi

36. An irrigation canal must supply 1060 ft^3/sec of water at a uniform depth of 5 ft. The canal has a 10 ft wide, rectangular cross section and is constructed of brick to make use of local building materials. The slope of the canal that will meet the supply conditions is most nearly

(A) 1.0%

(B) 1.5%

(C) 2.3%

(D) 3.4%

37. A small basin consists of the cover types given in the following table.

unit	area (ft^2)	cover type
1	200,000	open space with 80% grass, high infiltration
2	300,000	residential, 1/3 ac, moderate infiltration
3	180,000	paved roads and parking

According to the NRCS method, the soil storage capacity is most nearly

(A) 1.0 in

(B) 2.5 in

(C) 3.0 in

(D) 4.5 in

38. A plain sedimentation tank removes 100% of a sandy material with a mean specific gravity of 2.2, a mean diameter of 6.5×10^{-5} ft, and an operating temperature of 90°F. The system has a detention time of 2.5 hr and a flow of 18 ft^3/sec. The area and depth of the tank, respectively, are most nearly

(A) 10,000 ft^2; 13 ft

(B) 12,000 ft^2; 7 ft

(C) 14,000 ft^2; 16 ft

(D) 16,000 ft^2; 10 ft

39. Wastewater design flow for a wastewater treatment plant is to be based on population for domestic sewage, plus industrial wastewater, storm water, and infiltration. The parameters are given in the following table.

parameter	value
population	65,000 people
industrial area	200 ac
peak daily flow	200% of annual average
excess infiltration rate (above normal)	600 gal/day-in-mi
collection system properties	5 mi of 24 in pipe
	16 mi of 12 in pipe
	22 mi of 8 in pipe
industrial waste	10,000 gal/ac-day

The design maximum daily flow is most nearly

(A) 10 MGD

(B) 15 MGD

(C) 20 MGD

(D) 25 MGD

40. Sulfur dioxide is to be used to dechlorinate an effluent containing a chlorine residual of 6 mg/L as Cl_2. The design flow is 80 L/s. The amount of sulfur dioxide required is most nearly

(A) 10 kg/d

(B) 20 kg/d

(C) 40 kg/d

(D) 60 kg/d

STOP!

DO NOT CONTINUE!

This concludes the Morning Session of the examination. If you finish early, check your work and make sure that you have followed all instructions. After checking your answers, you may turn in your examination booklet and answer sheet and leave the examination room. Once you leave, you will not be permitted to return to work or change your answers.

Afternoon Session

In accordance with the rules established by your state, you may use textbooks, handbooks, bound reference materials, and any approved battery- or solar-powered, silent calculator to work this examination. However, no blank papers, writing tablets, unbound scratch paper, or loose notes are permitted. Sufficient room for scratch work is provided in the Examination Booklet.

You are not permitted to share or exchange materials with other examinees. However, the books and other resources used in this afternoon session do not have to be the same as were used in the morning session.

The Examination Booklet for the module you selected when registering will be provided. You may not change the module you selected.

You will have four hours in which to work this session of the examination. Your score will be determined by the number of questions that you answer correctly. There is a total of 40 questions. All 40 questions must be worked correctly in order to receive full credit on the exam. There are no optional questions. Each question is worth 1 point. The maximum possible score for this section of the examination is 40 points.

Partial credit is not available. No credit will be given for methodology, assumptions, or work written in your Examination Booklet.

Record all of your answers on the Answer Sheet. No credit will be given for answers marked in the Examination Booklet. Mark your answers with the official examination pencil provided to you. Marks must be dark and must completely fill the bubbles. Record only one answer per question. If you mark more than one answer, you will not receive credit for the question. If you change an answer, be sure the old bubble is erased completely; incomplete erasures may be misinterpreted as answers.

If you finish early, check your work and make sure that you have followed all instructions. After checking your answers, you may turn in your Examination Booklet and Answer Sheet and leave the examination room. Once you leave, you will not be permitted to return to work or change your answers.

When permission has been given by your proctor, break the seal on the Examination Booklet. Check that all pages are present and legible. If any part of your Examination Booklet is missing, your proctor will issue you a new Booklet.

WAIT FOR PERMISSION TO BEGIN

Name: _____
       Last   First   Middle Initial

Examinee number: _____

Examination Booklet number: _____

Principles and Practice of Engineering Examination

Afternoon Session
Practice Examination

Depth Modules

Afternoon Session—Construction

41. Ⓐ Ⓑ Ⓒ Ⓓ
42. Ⓐ Ⓑ Ⓒ Ⓓ
43. Ⓐ Ⓑ Ⓒ Ⓓ
44. Ⓐ Ⓑ Ⓒ Ⓓ
45. Ⓐ Ⓑ Ⓒ Ⓓ
46. Ⓐ Ⓑ Ⓒ Ⓓ
47. Ⓐ Ⓑ Ⓒ Ⓓ
48. Ⓐ Ⓑ Ⓒ Ⓓ
49. Ⓐ Ⓑ Ⓒ Ⓓ
50. Ⓐ Ⓑ Ⓒ Ⓓ

51. Ⓐ Ⓑ Ⓒ Ⓓ
52. Ⓐ Ⓑ Ⓒ Ⓓ
53. Ⓐ Ⓑ Ⓒ Ⓓ
54. Ⓐ Ⓑ Ⓒ Ⓓ
55. Ⓐ Ⓑ Ⓒ Ⓓ
56. Ⓐ Ⓑ Ⓒ Ⓓ
57. Ⓐ Ⓑ Ⓒ Ⓓ
58. Ⓐ Ⓑ Ⓒ Ⓓ
59. Ⓐ Ⓑ Ⓒ Ⓓ
60. Ⓐ Ⓑ Ⓒ Ⓓ

61. Ⓐ Ⓑ Ⓒ Ⓓ
62. Ⓐ Ⓑ Ⓒ Ⓓ
63. Ⓐ Ⓑ Ⓒ Ⓓ
64. Ⓐ Ⓑ Ⓒ Ⓓ
65. Ⓐ Ⓑ Ⓒ Ⓓ
66. Ⓐ Ⓑ Ⓒ Ⓓ
67. Ⓐ Ⓑ Ⓒ Ⓓ
68. Ⓐ Ⓑ Ⓒ Ⓓ
69. Ⓐ Ⓑ Ⓒ Ⓓ
70. Ⓐ Ⓑ Ⓒ Ⓓ

71. Ⓐ Ⓑ Ⓒ Ⓓ
72. Ⓐ Ⓑ Ⓒ Ⓓ
73. Ⓐ Ⓑ Ⓒ Ⓓ
74. Ⓐ Ⓑ Ⓒ Ⓓ
75. Ⓐ Ⓑ Ⓒ Ⓓ
76. Ⓐ Ⓑ Ⓒ Ⓓ
77. Ⓐ Ⓑ Ⓒ Ⓓ
78. Ⓐ Ⓑ Ⓒ Ⓓ
79. Ⓐ Ⓑ Ⓒ Ⓓ
80. Ⓐ Ⓑ Ⓒ Ⓓ

Afternoon Session—Geotechnical

81. Ⓐ Ⓑ Ⓒ Ⓓ
82. Ⓐ Ⓑ Ⓒ Ⓓ
83. Ⓐ Ⓑ Ⓒ Ⓓ
84. Ⓐ Ⓑ Ⓒ Ⓓ
85. Ⓐ Ⓑ Ⓒ Ⓓ
86. Ⓐ Ⓑ Ⓒ Ⓓ
87. Ⓐ Ⓑ Ⓒ Ⓓ
88. Ⓐ Ⓑ Ⓒ Ⓓ
89. Ⓐ Ⓑ Ⓒ Ⓓ
90. Ⓐ Ⓑ Ⓒ Ⓓ

91. Ⓐ Ⓑ Ⓒ Ⓓ
92. Ⓐ Ⓑ Ⓒ Ⓓ
93. Ⓐ Ⓑ Ⓒ Ⓓ
94. Ⓐ Ⓑ Ⓒ Ⓓ
95. Ⓐ Ⓑ Ⓒ Ⓓ
96. Ⓐ Ⓑ Ⓒ Ⓓ
97. Ⓐ Ⓑ Ⓒ Ⓓ
98. Ⓐ Ⓑ Ⓒ Ⓓ
99. Ⓐ Ⓑ Ⓒ Ⓓ
100. Ⓐ Ⓑ Ⓒ Ⓓ

101. Ⓐ Ⓑ Ⓒ Ⓓ
102. Ⓐ Ⓑ Ⓒ Ⓓ
103. Ⓐ Ⓑ Ⓒ Ⓓ
104. Ⓐ Ⓑ Ⓒ Ⓓ
105. Ⓐ Ⓑ Ⓒ Ⓓ
106. Ⓐ Ⓑ Ⓒ Ⓓ
107. Ⓐ Ⓑ Ⓒ Ⓓ
108. Ⓐ Ⓑ Ⓒ Ⓓ
109. Ⓐ Ⓑ Ⓒ Ⓓ
110. Ⓐ Ⓑ Ⓒ Ⓓ

111. Ⓐ Ⓑ Ⓒ Ⓓ
112. Ⓐ Ⓑ Ⓒ Ⓓ
113. Ⓐ Ⓑ Ⓒ Ⓓ
114. Ⓐ Ⓑ Ⓒ Ⓓ
115. Ⓐ Ⓑ Ⓒ Ⓓ
116. Ⓐ Ⓑ Ⓒ Ⓓ
117. Ⓐ Ⓑ Ⓒ Ⓓ
118. Ⓐ Ⓑ Ⓒ Ⓓ
119. Ⓐ Ⓑ Ⓒ Ⓓ
120. Ⓐ Ⓑ Ⓒ Ⓓ

Afternoon Session—Structural

121. Ⓐ Ⓑ Ⓒ Ⓓ
122. Ⓐ Ⓑ Ⓒ Ⓓ
123. Ⓐ Ⓑ Ⓒ Ⓓ
124. Ⓐ Ⓑ Ⓒ Ⓓ
125. Ⓐ Ⓑ Ⓒ Ⓓ
126. Ⓐ Ⓑ Ⓒ Ⓓ
127. Ⓐ Ⓑ Ⓒ Ⓓ
128. Ⓐ Ⓑ Ⓒ Ⓓ
129. Ⓐ Ⓑ Ⓒ Ⓓ
130. Ⓐ Ⓑ Ⓒ Ⓓ

131. Ⓐ Ⓑ Ⓒ Ⓓ
132. Ⓐ Ⓑ Ⓒ Ⓓ
133. Ⓐ Ⓑ Ⓒ Ⓓ
134. Ⓐ Ⓑ Ⓒ Ⓓ
135. Ⓐ Ⓑ Ⓒ Ⓓ
136. Ⓐ Ⓑ Ⓒ Ⓓ
137. Ⓐ Ⓑ Ⓒ Ⓓ
138. Ⓐ Ⓑ Ⓒ Ⓓ
139. Ⓐ Ⓑ Ⓒ Ⓓ
140. Ⓐ Ⓑ Ⓒ Ⓓ

141. Ⓐ Ⓑ Ⓒ Ⓓ
142. Ⓐ Ⓑ Ⓒ Ⓓ
143. Ⓐ Ⓑ Ⓒ Ⓓ
144. Ⓐ Ⓑ Ⓒ Ⓓ
145. Ⓐ Ⓑ Ⓒ Ⓓ
146. Ⓐ Ⓑ Ⓒ Ⓓ
147. Ⓐ Ⓑ Ⓒ Ⓓ
148. Ⓐ Ⓑ Ⓒ Ⓓ
149. Ⓐ Ⓑ Ⓒ Ⓓ
150. Ⓐ Ⓑ Ⓒ Ⓓ

151. Ⓐ Ⓑ Ⓒ Ⓓ
152. Ⓐ Ⓑ Ⓒ Ⓓ
153. Ⓐ Ⓑ Ⓒ Ⓓ
154. Ⓐ Ⓑ Ⓒ Ⓓ
155. Ⓐ Ⓑ Ⓒ Ⓓ
156. Ⓐ Ⓑ Ⓒ Ⓓ
157. Ⓐ Ⓑ Ⓒ Ⓓ
158. Ⓐ Ⓑ Ⓒ Ⓓ
159. Ⓐ Ⓑ Ⓒ Ⓓ
160. Ⓐ Ⓑ Ⓒ Ⓓ

Afternoon Session—Transportation

161. Ⓐ Ⓑ Ⓒ Ⓓ   171. Ⓐ Ⓑ Ⓒ Ⓓ   181. Ⓐ Ⓑ Ⓒ Ⓓ   191. Ⓐ Ⓑ Ⓒ Ⓓ
162. Ⓐ Ⓑ Ⓒ Ⓓ   172. Ⓐ Ⓑ Ⓒ Ⓓ   182. Ⓐ Ⓑ Ⓒ Ⓓ   192. Ⓐ Ⓑ Ⓒ Ⓓ
163. Ⓐ Ⓑ Ⓒ Ⓓ   173. Ⓐ Ⓑ Ⓒ Ⓓ   183. Ⓐ Ⓑ Ⓒ Ⓓ   193. Ⓐ Ⓑ Ⓒ Ⓓ
164. Ⓐ Ⓑ Ⓒ Ⓓ   174. Ⓐ Ⓑ Ⓒ Ⓓ   184. Ⓐ Ⓑ Ⓒ Ⓓ   194. Ⓐ Ⓑ Ⓒ Ⓓ
165. Ⓐ Ⓑ Ⓒ Ⓓ   175. Ⓐ Ⓑ Ⓒ Ⓓ   185. Ⓐ Ⓑ Ⓒ Ⓓ   195. Ⓐ Ⓑ Ⓒ Ⓓ
166. Ⓐ Ⓑ Ⓒ Ⓓ   176. Ⓐ Ⓑ Ⓒ Ⓓ   186. Ⓐ Ⓑ Ⓒ Ⓓ   196. Ⓐ Ⓑ Ⓒ Ⓓ
167. Ⓐ Ⓑ Ⓒ Ⓓ   177. Ⓐ Ⓑ Ⓒ Ⓓ   187. Ⓐ Ⓑ Ⓒ Ⓓ   197. Ⓐ Ⓑ Ⓒ Ⓓ
168. Ⓐ Ⓑ Ⓒ Ⓓ   178. Ⓐ Ⓑ Ⓒ Ⓓ   188. Ⓐ Ⓑ Ⓒ Ⓓ   198. Ⓐ Ⓑ Ⓒ Ⓓ
169. Ⓐ Ⓑ Ⓒ Ⓓ   179. Ⓐ Ⓑ Ⓒ Ⓓ   189. Ⓐ Ⓑ Ⓒ Ⓓ   199. Ⓐ Ⓑ Ⓒ Ⓓ
170. Ⓐ Ⓑ Ⓒ Ⓓ   180. Ⓐ Ⓑ Ⓒ Ⓓ   190. Ⓐ Ⓑ Ⓒ Ⓓ   200. Ⓐ Ⓑ Ⓒ Ⓓ

Afternoon Session—Water Resources and Environmental

201. Ⓐ Ⓑ Ⓒ Ⓓ   211. Ⓐ Ⓑ Ⓒ Ⓓ   221. Ⓐ Ⓑ Ⓒ Ⓓ   231. Ⓐ Ⓑ Ⓒ Ⓓ
202. Ⓐ Ⓑ Ⓒ Ⓓ   212. Ⓐ Ⓑ Ⓒ Ⓓ   222. Ⓐ Ⓑ Ⓒ Ⓓ   232. Ⓐ Ⓑ Ⓒ Ⓓ
203. Ⓐ Ⓑ Ⓒ Ⓓ   213. Ⓐ Ⓑ Ⓒ Ⓓ   223. Ⓐ Ⓑ Ⓒ Ⓓ   233. Ⓐ Ⓑ Ⓒ Ⓓ
204. Ⓐ Ⓑ Ⓒ Ⓓ   214. Ⓐ Ⓑ Ⓒ Ⓓ   224. Ⓐ Ⓑ Ⓒ Ⓓ   234. Ⓐ Ⓑ Ⓒ Ⓓ
205. Ⓐ Ⓑ Ⓒ Ⓓ   215. Ⓐ Ⓑ Ⓒ Ⓓ   225. Ⓐ Ⓑ Ⓒ Ⓓ   235. Ⓐ Ⓑ Ⓒ Ⓓ
206. Ⓐ Ⓑ Ⓒ Ⓓ   216. Ⓐ Ⓑ Ⓒ Ⓓ   226. Ⓐ Ⓑ Ⓒ Ⓓ   236. Ⓐ Ⓑ Ⓒ Ⓓ
207. Ⓐ Ⓑ Ⓒ Ⓓ   217. Ⓐ Ⓑ Ⓒ Ⓓ   227. Ⓐ Ⓑ Ⓒ Ⓓ   237. Ⓐ Ⓑ Ⓒ Ⓓ
208. Ⓐ Ⓑ Ⓒ Ⓓ   218. Ⓐ Ⓑ Ⓒ Ⓓ   228. Ⓐ Ⓑ Ⓒ Ⓓ   238. Ⓐ Ⓑ Ⓒ Ⓓ
209. Ⓐ Ⓑ Ⓒ Ⓓ   219. Ⓐ Ⓑ Ⓒ Ⓓ   229. Ⓐ Ⓑ Ⓒ Ⓓ   239. Ⓐ Ⓑ Ⓒ Ⓓ
210. Ⓐ Ⓑ Ⓒ Ⓓ   220. Ⓐ Ⓑ Ⓒ Ⓓ   230. Ⓐ Ⓑ Ⓒ Ⓓ   240. Ⓐ Ⓑ Ⓒ Ⓓ

Afternoon Session Construction

41. Consider the two cross sections shown.

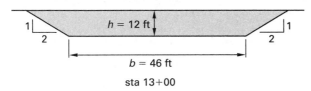

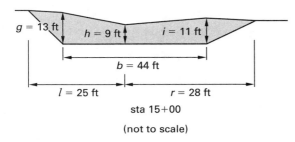

(not to scale)

Using the average end area method, what is most nearly the volume between sta 13+00 and sta 15+00?

(A) 3800 yd^3

(B) 4200 yd^3

(C) 4600 yd^3

(D) 5000 yd^3

42. A planimeter recorded the areas enclosed by the 700 ft and 707 ft grading contours as 1200 ft^2 and 1800 ft^2, respectively.

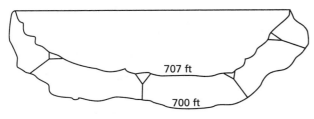

The volume of material enclosed by the contour lines is most nearly

(A) 4000 ft^3

(B) 11,000 ft^3

(C) 14,000 ft^3

(D) 21,000 ft^3

43. The following table shows unreduced profile leveling readings.

station	BS	HI	FS	IFS	elevation
BM 145	3.57				617.24
1+00				4.29	
2+00				5.17	
TP−1	2.65		8.21		
3+90				3.75	

The elevation at sta 3+90 is most nearly

(A) 609 ft

(B) 610 ft

(C) 612 ft

(D) 615 ft

44. A foundation contractor bid on constructing the residence shown. The foundation consists of exterior footings totaling 207 linear feet (measuring 16 in × 8 in) and four column footings (measuring 3 ft × 3 ft × 12 in).

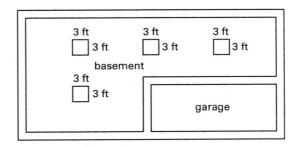

Due to poor soil conditions, the area of both the exterior and column footings must be increased by 50%, with no change to footing depths. The additional concrete needed to make the larger footings is most nearly

(A) 2.7 yd^3

(B) 4.1 yd^3

(C) 7.4 yd^3

(D) 12 yd^3

45. Consider the mass diagram shown.

Select the ground profile that best corresponds to the mass diagram.

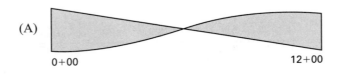

(A)

(B)

(C)

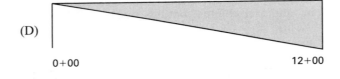

(D)

46. Skyline Road is constructed as follows.

- 6 in AC pavement at 110 lbf/yd² per inch of thickness
- 6 in AB pavement at 3000 lbf/yd³
- pavement width of 40 ft (both lanes)

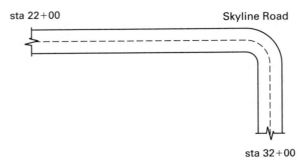

The amounts of AC and AB pavement, respectively, from sta 22+00 to sta 32+00 is most nearly

(A) 1470 tons and 1110 tons

(B) 2560 tons and 1560 tons

(C) 3330 tons and 2220 tons

(D) 4400 tons and 3330 tons

47. A contractor is bidding on a painting job with the following considerations.

- The total area to be painted measures 34,600 ft².
- A five-gallon bucket of paint costs $10.
- A five-gallon bucket of paint covers 1200 ft².
- Labor costs $25/hr.
- The production rate is 700 ft²/hr.
- The overhead cost of labor is 35%.
- The overhead cost of materials is 30%.

The total bid for the job is most nearly

(A) $1850

(B) $1920

(C) $2040

(D) $3710

48. A building with 16 bays is to be constructed with the following considerations.

- Bay type A will cost $100,000.
- Bay type B will cost $85,000.
- Bay type C will cost $70,000.
- The total cost of entrances will be $20,000.
- An elevator will cost $40,000.
- The total cost of end walls will be $60,000.
- All equipment will cost $80,000.

A	C	C	A
B	A	B	B
B	B	A	B
A	C	C	A

Using the bay method of cost estimating, the cost of building construction will most nearly be

- (A) $1,190,000
- (B) $1,390,000
- (C) $1,400,000
- (D) $1,590,000

49. A civil engineering firm purchases computers for $30,000. IRS rules allow a five year useful life. The total salvage value is $2500 at the end of the five year useful life. Using the straight-line method, the depreciation amount the firm can take annually is most nearly

- (A) $2500
- (B) $4000
- (C) $5500
- (D) $6000

50. A civil engineering firm purchases computers for $30,000. IRS rules allow a five year useful life. The total salvage value is $2500 at the end of the five year useful life. The owner decides to use the sum of the years' digits depreciation method. The total depreciation at the end of the second year is most nearly

- (A) $6500
- (B) $8300
- (C) $17,000
- (D) $21,000

51. EBITDA is a measure of a company's

- (A) financial performance
- (B) worth prior to going public
- (C) free cash flow
- (D) profit

52. An excavation in type A soil will be open for less than 24 hours. If a simple slope with a height of 12 ft is desired, the maximum slope (H:V) allowed by OSHA is

- (A) $^{1}/_{2}$:1
- (B) $^{3}/_{4}$:1
- (C) 1:1
- (D) 2:1

53. Experience modification (X-mod) is a

- (A) rating used to reduce a firm's umbrella liability policy
- (B) rating that allows credit for a contractor's extended work history
- (C) factor used to determine errors and omissions in deductible amounts
- (D) factor that compares an insured's actual losses to average losses in the same industry and period

54. A heavy-construction contractor reports a total of five injuries and illnesses in a period of 80,000 working hours. The firm's incident rate as defined by the U.S. Department of Labor, Bureau of Labor Statistics, is most nearly

- (A) 3
- (B) 6
- (C) 13
- (D) 18

55. The following illustration depicts the results of a foundation load test carried out over a 60 hour period. The tested RC pile has a diameter of 2 ft and is 70 ft long.

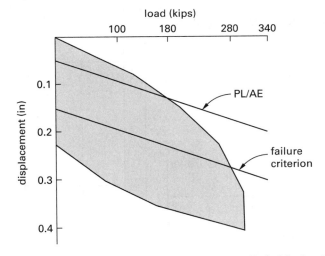

If the desired factor of safety is 2.0, the allowable load per pile is most nearly

- (A) 90 kips
- (B) 140 kips
- (C) 180 kips
- (D) 280 kips

56. A retaining wall is built to support soil with an angle of internal friction of 33°.

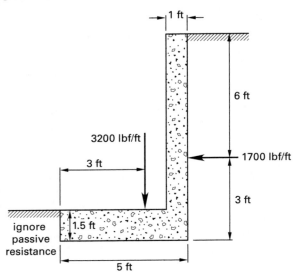

The factor of safety against overturning is most nearly

(A) 1.7

(B) 1.8

(C) 1.9

(D) 2.0

57. Consider the retaining wall shown. The soil's angle of internal friction is 33°.

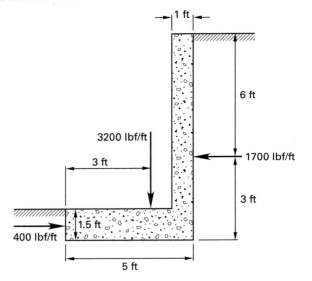

Without consideration for adhesion at the footing base, the factor of safety against sliding is most nearly

(A) 1.0

(B) 1.3

(C) 1.6

(D) 1.9

58. Consider the framing and structural details shown.

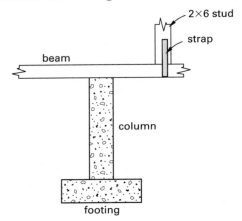

The load path for this design is

(A) stud → beam → strap → column → footing → soil

(B) beam → stud → strap → column → footing → soil

(C) soil → footing → column → beam → stud → strap

(D) stud → strap → beam → column → footing → soil

59. The new director of a geotechnical lab discovers that a technician has been rolling soils to a $1/16$ in diameter thread instead of the prescribed $1/8$ in diameter thread when performing the plastic limit test. Taking the plasticity index (PI), the liquid limit (LL), and the plastic limit into account, this action would

(A) decrease the PI value as tested by ASTM D4318

(B) increase the PI value as tested by ASTM D4318

(C) decrease the LL but increase the PL values

(D) increase both the LL and PI values, thereby cancelling each other out

60. A designer of industrial floor slabs is given the following data.

• The water/cement ratio must be less than 0.45.

• In 28 days, the compressive strength must reach 4500 psi.

• The percentage of soluble sulfates (SO_4) in the soil is 1.5%.

The most appropriate cement to specify in the concrete mix is

(A) type IA

(B) type II

(C) type IIIA

(D) type V

61. Which of the following is NOT an OSHA soil classification?

(A) rock

(B) layered geological strata

(C) type C-60

(D) type A

62. SMAW, GMAW, SAW, and FCAW methods are associated with

(A) post-tensioning wire strand shapes

(B) welding techniques

(C) methods of welded wire fabric slab design

(D) wastewater quality standards

63. A Caterpillar 825H sheepsfoot roller compacts a 9 in soil lift at a speed of 7 mph and does five passes. Each of the roller's four wheels is 3.7 ft wide. The production rate (in compacted yd^3/hr) is most nearly

(A) 580 yd^3/hr

(B) 680 yd^3/hr

(C) 760 yd^3/hr

(D) 1500 yd^3/hr

64. A smooth vibratory drum roller has the assumed and actual production characteristics given in the following table.

	assumed	actual
thickness	5 in	4 in
speed	6 mph	5 mph
no. of passes	7 per hour	6 per hour
production	480 yd^3/hr	

Assuming actual compaction meets the methods-only compaction specifications, the actual production is most nearly

(A) 270 yd^3/hr

(B) 370 yd^3/hr

(C) 540 yd^3/hr

(D) 590 yd^3/hr

65. A mining truck has a 120 ton gross machine weight (GMW). An access road has a 7% grade and 3% rolling resistance. If the truck is moving at a constant speed of 10 mph uphill, the required horsepower is most nearly

(A) 640 hp

(B) 960 hp

(C) 1300 hp

(D) 1600 hp

66. A Rimpull scraper with a 300 hp engine is operating at 80% efficiency and 10 mph. The machine rim pull is most nearly

(A) 8000 lbf

(B) 9000 lbf

(C) 10,000 lbf

(D) 11,000 lbf

67. When making dewatering calculations, the most critical aspect affecting dewatering time is the

(A) header pipe diameter

(B) radius of influence

(C) capacity of the well point system

(D) soil permeability

68. The lifting capacity of a crane

(A) increases with an increase in load radius

(B) decreases with an increase in load radius

(C) increases with an increase in boom length for a given load radius

(D) is unrelated to the load radius

69. A concrete floor slab for a hotel lobby is ready for bidding. The considerations for placing the bid are as follows.

- The lobby area measures 10,000 ft^2 (100 ft × 100 ft).

- In total, the graded base will cost $12,000.

- In total, the form work will cost $9000.

- A 6 in thick concrete slab costs $200/$yd^3$.

- No. 6 rebar at 9 in on center both ways costs $5/ linear foot.

- Vapor retarder costs $6/$ft^2$.

- Labor cost is $2/$ft^2$.

- Waste disposal will cost 7% of the total estimated cost.

The bid amount is most nearly

(A) $238,000

(B) $255,000

(C) $273,000

(D) $291,000

70. The following information is gathered during a project's take-off analysis.

steel shape	length (ft)	count	cost
W10 × 22	350	3	$7/lbm
W18 × 65	460	4	$10/lbm
W24 × 76	700	2	$12/lbm

The estimated cost of structural steel for the project is

(A) $2,635,000

(B) $2,724,000

(C) $2,806,000

(D) $2,893,000

71. The initial cost of a plotter is $30,000. Maintenance, salvage, and equivalent uniform annual cost are as shown.

year	maintenance cost ($)	salvage value ($)	EUAC ($)
1	2000	10,000	16,000
2	3000	9000	15,000
3	4000	5000	17,000
4	7000	4000	18,000
5	8000	3000	23,000

The plotter should be replaced at the end of year

(A) 2

(B) 3

(C) 4

(D) 5

72. A credit union building has the following characteristics.

- The building area is 50 ft × 80 ft.
- The basement excavation depth is 10 ft.
- The stable slopes are at $^1/_2$:1 (H:V) on all sides.

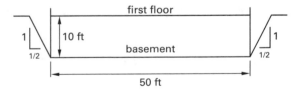

The excavated volume is most nearly

(A) 1500 yd³

(B) 1700 yd³

(C) 2000 yd³

(D) 47,000 yd³

73. After three years of investing at an effective annual interest rate of 8%, a company's accumulation is exactly $50,000. During those three years, the average inflation rate was 4%. The original principal invested three years ago was most nearly

(A) $34,100

(B) $35,300

(C) $36,500

(D) $39,700

74. Your 401(k) is presently worth $10,000. The average return from the stock market is 12% per year. At that rate, the value of your 401(k) will double and triple, respectively, in approximately

(A) 4 yr, 8 yr

(B) 4 yr, 10 yr

(C) 6 yr, 8 yr

(D) 6 yr, 10 yr

75. A construction consultant estimates the following productivity rates.

activity	quantity	productivity
earthwork	600 hr	12 hr/day
surveying	60 hr	10 hr/day
design phase	500 hr	10 hr/day
permitting	100 hr	5 hr/day

The total activity duration is most nearly

(A) 32 days

(B) 130 days

(C) 780 days

(D) 1200 days

76. A project has the following performance measures.

$$\text{actual cost of work performed (ACWP)} = \$380,000$$
$$\text{cost variance (CV)} = \$20,000$$
$$\text{budgeted cost of work performed (BCWP)} = \$400,000$$
$$\text{percent cost variance (PTV)} = 6\%$$

The percent cost variable (PCV) is most nearly

(A) −6%

(B) −5%

(C) +5%

(D) +6%

77. A project has the following activity graph.

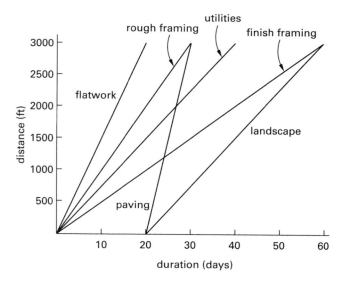

For the owners to see the greatest number of activities during their site visit, they should visit between days

(A) 0 and 10

(B) 10 and 20

(C) 20 and 30

(D) 40 and 60

78. Parties 1 and 2 are negotiating a contract. The deadline to close the deal is in a few hours. Use the following Pareto optimal curve to schedule the negotiation.

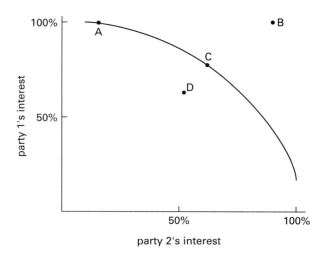

Considering only points A, B, C, and D, when is the best time for the two parties to reach an agreement?

(A) Point A, because at least one party is 100% interested.

(B) Point B, because both parties are equally and maximally interested.

(C) Point C, because both parties are equally and maximally interested.

(D) Point D, because both parties are equally interested.

79. In a career performance prediction test, the following four traits are evaluated.

- empathy: cares about employees

- optimism: has a great future outlook

- leadership: effectively delegates responsibility and motivates staff

- stress management: can multitask with overlapping deadlines

From most important to least important, what would be the ranking order of these traits for the ideal candidate to lead a 50 person engineering department?

(A) leadership, optimism, empathy, and stress management

(B) optimism, empathy, stress management, and leadership

(C) empathy, optimism, leadership, and stress management

(D) stress management, leadership, empathy, and optimism

80. A planned tunneling project is expected to take approximately 4000 hours to complete. The owner is not in a rush to complete construction. Given this information, the best of the proposed work schedules is

(A) 10 hr/day, 15 pieces of equipment, 50 people, 400 days

(B) 10 hr/day, 20 pieces of equipment, 50 people, 380 days

(C) 12 hr/day, 12 pieces of equipment, 80 people, 333 days

(D) 12 hr/day, 15 pieces of equipment, 60 people, 333 days

Afternoon Session Geotechnical

81. A rigid foundation is supported by friction piles in clay as shown in the following plan and elevation views. The total load on the piles, reduced by the displaced soil weight, is 2000 kN. The settlement of layer 2 is most nearly

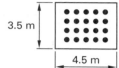

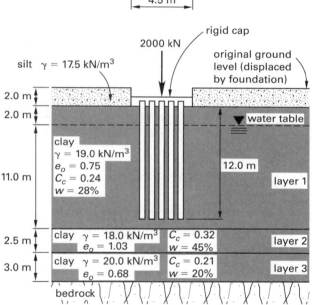

(A) 12 mm

(B) 17 mm

(C) 24 mm

(D) 35 mm

82. A 250 mm layer of soil bentonite will be placed just beneath the geomembrane liner of a proposed landfill. The soil bentonite layer will be placed in two 125 mm lifts, and the bentonite content will be 8% by dry weight. If the compacted moist unit weight of this soil bentonite layer is 17.0 kN/m^3 with a moisture content of 18%, the amount of dry bentonite that must be spread for mixing on each lift is most nearly

(A) 0.14 kN/m^2

(B) 0.18 kN/m^2

(C) 0.22 kN/m^2

(D) 0.26 kN/m^2

83. A landfill is 300 m × 400 m in plan. The clay liner has a hydraulic conductivity of 6×10^{-7} cm/s and experiences an average annual leachate head of 0.5 m as shown. A subgrade drain (pore pressure is atmospheric) lies below the clay liner. Assume one-dimensional flow downward for the leachate. The annual steady-state flow rate from this landfill is most nearly

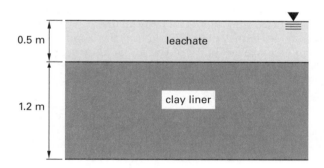

(A) 10^3 m^3/year

(B) 10^4 m^3/year

(C) 10^5 m^3/year

(D) 10^6 m^3/year

84. The groundwater table on a project site (elevation view shown) will be lowered 16.3 m. The groundwater table is now at the ground surface. Assume a soil moisture content of 11% above the groundwater table once it is lowered. After lowering, the settlement of the clay layer will be most nearly

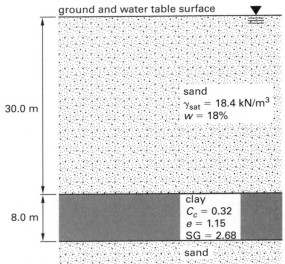

(A) 0.10 m

(B) 0.20 m

(C) 0.30 m

(D) 0.40 m

85. The permeability of a soil is evaluated in a falling-head permeameter. The head decreases from 100 cm to 50 cm in 21 min 18 s. The body diameter is 10 cm, the standpipe diameter is 0.25 cm, and the sample length is 6 cm. The permeability of the soil is most nearly

(A) 1×10^{-7} cm/s

(B) 2×10^{-7} cm/s

(C) 1×10^{-6} cm/s

(D) 2×10^{-6} cm/s

86. A site consists of 25 m of clayey silt that is to be consolidated for eventual placement of a large office building. From a consolidation test with a soil sample 5.0 cm high, it has been determined that the time to achieve 90% consolidation (of the soil sample) is 10 min 46 s. Assuming double drainage for both the sample and the clayey silt layer, how much time would be required to achieve 90% consolidation of the 25 m clayey silt layer?

(A) 1 yr

(B) 2 yr

(C) 5 yr

(D) 15 yr

87. A clay layer 10 m thick (with double drainage) is expected to have an ultimate settlement of 502 mm. If the settlement in 5 yr is 124 mm, the remaining time it will take to reach a settlement of 250 mm is most nearly

(A) 5 yr

(B) 10 yr

(C) 15 yr

(D) 20 yr

88. A smooth retaining wall holds back a uniform sand backfill as shown. The distance of the active resultant force from the bottom of the retaining wall is most nearly

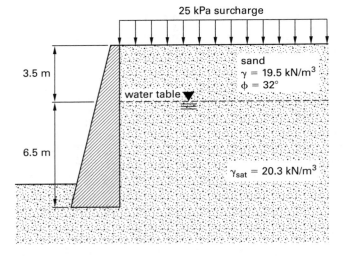

(A) 2.9 m

(B) 3.3 m

(C) 3.8 m

(D) 4.5 m

89. A loose, natural sand deposit has a saturated unit weight of 19.3 kN/m³ and an angle of internal friction of 29°. The water table is at the ground surface. The total at-rest lateral earth pressure at a depth of 10 m is most nearly

(A) 80 kPa

(B) 150 kPa

(C) 210 kPa

(D) 240 kPa

90. An artificial reservoir holds a constant level of water as shown. A compacted clay liner with the given properties is used to contain the water. Porous sand underlays the clay liner. The true water velocity (pore velocity) through the clay liner is most nearly

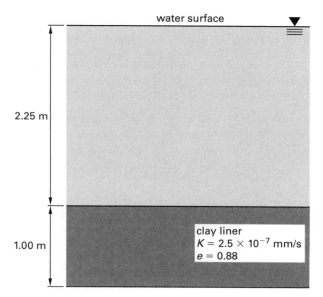

(A) 5.6×10^{-7} mm/s

(B) 6.7×10^{-7} mm/s

(C) 1.2×10^{-6} mm/s

(D) 3.1×10^{-6} mm/s

91. A smooth gravity retaining wall holds soil backfill with properties as shown. Disregard passive earth pressure. The vertical pressure at point A is most nearly

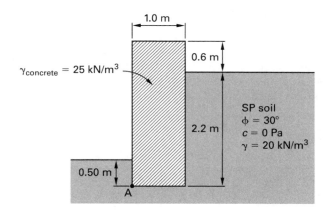

(A) 100 kPa

(B) 120 kPa

(C) 125 kPa

(D) 140 kPa

92. A sample of saturated clay has a total mass of 1733 g and a dry mass of 1287 g. The specific gravity of the soil particles is 2.7. The total unit weight of this soil is most nearly

(A) 17.1 kN/m^3

(B) 17.7 kN/m^3

(C) 18.0 kN/m^3

(D) 18.4 kN/m^3

93. A medium uniform sand has the gradation shown. The sand has a dry unit weight of 15.8 kN/m^3, and the particles have a specific gravity of 2.65.

sieve no.	sieve size (mm)	percent finer (by mass)
10	2.00	100.0
20	0.850	99.7
30	0.600	93.0
40	0.425	58.2
50	0.300	42.9
70	0.212	18.2
100	0.150	10.1
200	0.075	1.0

The estimated coefficient of permeability for this sand is most nearly

(A) 2×10^{-3} cm/s

(B) 2×10^{-2} cm/s

(C) 2 cm/s

(D) 20 cm/s

94. A concrete dam impounds water. Using the flow net shown, the pore pressure at point A is most nearly

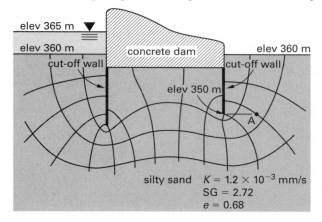

(A) 80 kPa

(B) 105 kPa

(C) 125 kPa

(D) 140 kPa

95. The soil profile and the properties of each soil layer beneath a reservoir are shown. The sandy layer at the bottom of the soil profile has horizontal drainage and zero pore pressure. The water level of the reservoir is constant, and the total area of the reservoir is 5000 m². Assuming vertical flow through the soil profile, the seepage water loss from the reservoir in six months is most nearly

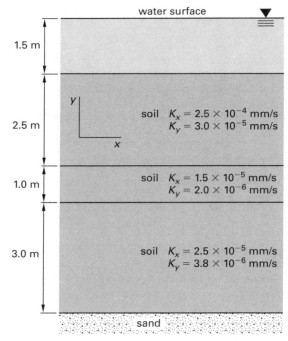

(A) 85 m³

(B) 94 m³

(C) 1000 m³

(D) 1200 m³

96. What is the effective area of the rectangular footing supporting a concentrated normal force as shown?

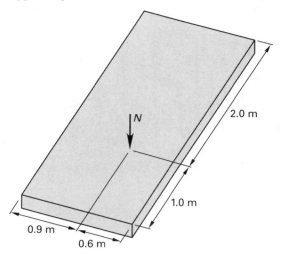

(A) 0.6 m²

(B) 1.8 m²

(C) 2.4 m²

(D) 4.5 m²

97. A long wall footing that is 2 m wide is situated on stiff, saturated clay. The depth of the footing is 1 m. The clay has a unit weight of 18.5 kN/m³ and an undrained shear strength of 110 kPa. Loading is applied rapidly enough that undrained conditions prevail ($\phi = 0$).

Use Terzaghi bearing capacity factors and the following bearing capacity formula.

$$q_{\text{ult}} = c\lambda_{cs}\lambda_{cd}N_c + q\lambda_{qs}\lambda_{qd}N_q + \tfrac{1}{2}\lambda_{\gamma s}\lambda_{\gamma d}\gamma B N_\gamma$$

The shape and depth factors are

$$\lambda_{qs} = \lambda_{\gamma s} = 1$$
$$\lambda_{qd} = \lambda_{\gamma d} = 1$$
$$\lambda_{cs} = 1 + 0.2\left(\frac{B}{L}\right)\tan^2\left(45° + \frac{\phi}{2}\right)$$
$$\lambda_{cd} = 1 + 0.2\left(\frac{D_f}{B}\right)\tan\left(45° + \frac{\phi}{2}\right)$$

The ultimate bearing capacity per meter length of footing is most nearly

(A) 300 kN/m

(B) 600 kN/m

(C) 1000 kN/m

(D) 1400 kN/m

98. A rock core is retrieved from a drill hole. The length of the recovered core is 123 cm. There are five pieces 10 cm or more in length, and the pieces have a combined length of 89 cm.

The rock quality designation for this core is most nearly

(A) 0%

(B) 38%

(C) 72%

(D) 138%

99. A 0.30 m diameter prestressed concrete pile has been driven 6 m into a dense sand deposit. The soil-pile friction angle is 25°. The unit weight of the prestressed concrete pile is 25 kN/m³, and the unit weight of the sand is 20 kN/m³. Assume that the critical depth is 20 times the diameter of the pile, and that the horizontal earth pressure coefficient for tension is 1.1. The ultimate pullout load capacity of the pile is most nearly

(A) 160 kN

(B) 170 kN

(C) 180 kN

(D) 190 kN

100. A soil profile has the properties shown. The average permanent vertical pressure on the normally consolidated clay layer is expected to increase by 130 kPa. The average effective overburden pressure at the middle of the clay layer is 240 kPa. The total primary consolidation settlement is most nearly

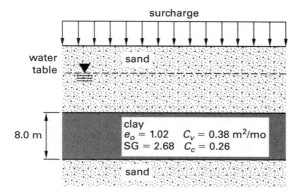

(A) 100 mm

(B) 180 mm

(C) 190 mm

(D) 200 mm

101. The clay soil shown undergoes consolidation. The percent consolidation at mid-depth of the clay three years after loading is most nearly

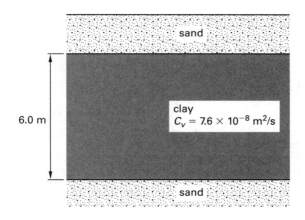

(A) 30%

(B) 40%

(C) 70%

(D) 90%

102. A double-drained clay layer 20 m thick settles 18.2 cm in 5 yr. The coefficient of consolidation for this clay is 4.3×10^{-7} m^2/s. The time required for the clay layer to undergo 90% of its ultimate primary consolidation settlement amount is most nearly

(A) 4.2 yr

(B) 5.8 yr

(C) 6.3 yr

(D) 7.2 yr

103. A clay soil is loaded as shown and undergoes consolidation. If one-dimensional loading is assumed, what is most nearly the excess pore water pressure at mid-depth of the clay layer immediately after loading?

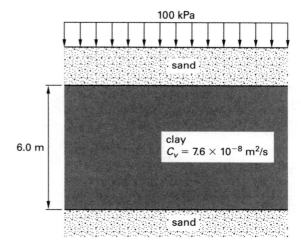

(A) 0 kPa

(B) 35 kPa

(C) 71 kPa

(D) 100 kPa

104. A soil has a wet unit weight of 17.6 kN/m^3 and a moisture content of 8%. The specific gravity of the solid particles is 2.72. The degree of saturation is most nearly

(A) 8%

(B) 14%

(C) 28%

(D) 34%

105. A braced cut in clay has properties and dimensions as shown in the illustration. The horizontal center-to-center spacing of the struts is 4.0 m. The load on the bottom strut (strut A) is most nearly

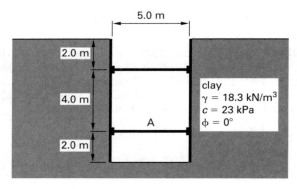

(A) 190 kN

(B) 240 kN

(C) 940 kN

(D) 1000 kN

106. A soil has a grain size distribution as shown in the sieve analysis chart. (See accompanying chart.) The coefficient of curvature for this soil is most nearly

(A) 2.6

(B) 3.5

(C) 4.3

(D) 4.5

107. A 10 m long precast concrete pile is installed into homogeneous sand. The pile cross section is 254 mm × 254 mm. The unit weight of the sand is 18.8 kN/m³. The internal friction angle of the sand is 35°. Consider the critical depth as 15 times the width of the pile. If the earth pressure coefficient is 1.6 and the soil-pile friction angle is 0.6ϕ, then the total frictional resistance of the pile in the sand is most nearly

(A) 80 kN

(B) 250 kN

(C) 360 kN

(D) 480 kN

108. A wall is supporting a horizontal force, P, as shown. Assume that the wall is smooth. The total force per meter of wall that the soil can sustain is most nearly

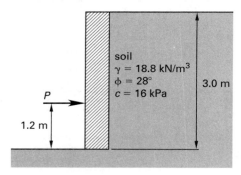

(A) 310 kN/m

(B) 400 kN/m

(C) 480 kN/m

(D) 500 kN/m

Chart for Problem 106

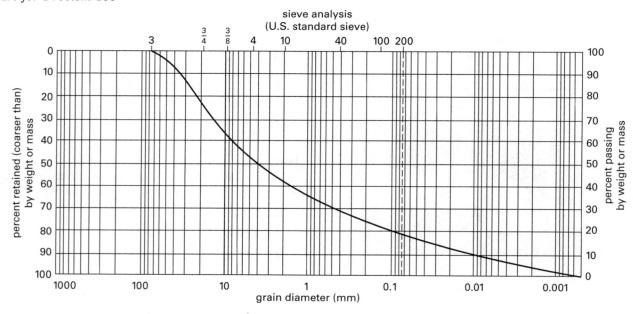

109. A consolidated-drained test (S-test) is performed on a sand sample. Initially, the saturated sand is consolidated in the triaxial cell under an equal all-around pressure of 200 kPa. Maintaining the cell pressure, the axial stress is increased 468 kPa. Under this stress state, the sample is at failure. The angle of internal friction of the sample is most nearly

(A) 0°

(B) 30°

(C) 33°

(D) 38°

110. A layer of sand has particles with a specific gravity of 2.66 and a void ratio of 0.62. The buoyant unit weight of the sand is most nearly

(A) 10 kN/m^3

(B) 11 kN/m^3

(C) 12 kN/m^3

(D) 20 kN/m^3

111. A continuous wall footing 1.5 m wide supports a load of 596 kN/m. The unit weight of the soil beneath the foundation is 18.6 kN/m^3. The soil has a cohesion of 14 kPa and an angle of internal friction of 25°. If the footing is placed near the ground surface, and if the Terzaghi bearing capacity factors and formula are used, the factor of safety against bearing capacity failure is most nearly

(A) 0.8

(B) 1.2

(C) 1.8

(D) 2.8

112. A dry sand sample is tested in a direct shear box with a normal stress of 100 kPa. Failure occurs at a shear stress of 63.4 kPa. The size of the tested sample is 6 cm × 6 cm × 3 cm (height). For a normal stress of 75 kPa, what shear force would be required to cause failure in the sample?

(A) 0.17 kN

(B) 0.37 kN

(C) 2.8 kN

(D) 48 kN

113. An unconfined-undrained compression test is conducted on a clay soil sample that had an initial height of 9.1 cm and an initial diameter of 4.0 cm. The axial load at failure is 0.43 kN, and the corresponding height is 8.67 cm. The undrained shear strength of this clay is most nearly

(A) 80 kPa

(B) 160 kPa

(C) 180 kPa

(D) 320 kPa

114. A consolidated-drained test is performed on a normally consolidated clay. The chamber confining pressure is 280 kPa, and the deviator stress at failure is 410 kPa. Assume that the normally consolidated clay has no drained cohesion ($c' = 0$). The shear stress on the failure plane is most nearly

(A) 190 kPa

(B) 300 kPa

(C) 380 kPa

(D) 580 kPa

115. A sand has a minimum void ratio of 0.41 and a maximum void ratio of 0.78. Its dry unit weight is 16.5 kN/m^3. If the specific gravity of the solids is 2.65, the relative density of this sand is most nearly

(A) 0.40

(B) 0.55

(C) 0.65

(D) 0.80

116. A soil has the following properties.

liquid limit	40
plasticity index	13
percent passing no. 10 sieve	98%
percent passing no. 40 sieve	87%
percent passing no. 200 sieve	45%

The AASHTO classification and group index number is most nearly

(A) A-5 (3)

(B) A-6 (1)

(C) A-6 (3)

(D) A-7-6 (1)

117. Classify a soil with the following characteristics using the Unified Soil Classification System (USCS).

liquid limit	55
plastic limit	20
uniformity coefficient, C_u	12
compression index, C_c	1.5
percent passing 1 in sieve	100%
percent passing no. 4 sieve	98%
percent passing no. 40 sieve	45%
percent passing no. 200 sieve	26%

(A) SC

(B) SW

(C) SP

(D) SM

118. The active pressure on the sheet pile wall shown is in equilibrium with the passive pressure and the anchor force. Assume that the sheet pile is smooth and that the resultant force acting on the passive side is horizontal and acting at point C as shown. The value of the force per meter length of wall is most nearly

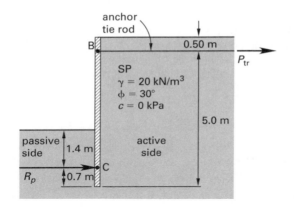

(A) 70 kN/m

(B) 80 kN/m

(C) 100 kN/m

(D) 120 kN/m

119. A mat foundation is 20 m × 31 m in plan. The total dead and live load is 33 540 kN. The depth needed for a fully compensated foundation is most nearly

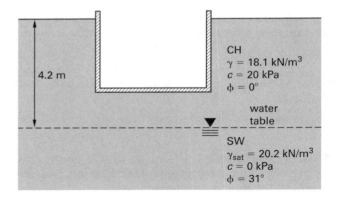

(A) 0.80 m

(B) 2.0 m

(C) 3.0 m

(D) 3.5 m

120. A concrete retaining wall has the specifications shown. If passive resistance is ignored, the factor of safety against sliding is most nearly

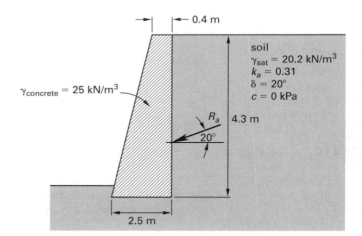

(A) 1.0

(B) 1.2

(C) 1.6

(D) 2.8

Afternoon Session
Structural

121. A two-lane highway bridge is constructed using precast concrete girders. The girders are simply supported and span 60 ft. The weight of girders and deck is such that the dead load bending moment at the critical location for bending moment is 500 ft-kips. If the bridge is designed for AASHTO HL-93 loading using LRFD strength criteria, only one lane is loaded at a time, and girder load distribution is not considered, the design bending moment per lane at the critical location is most nearly

(A) 2500 ft-kips

(B) 3000 ft-kips

(C) 3500 ft-kips

(D) 4000 ft-kips

122. A rigid diaphragm transfers a lateral wind force of 0.4 kip/ft into a system of shear walls whose relative rigidities, in multiples of R, against forces in the north direction are shown in the plan. The force in wall A of the system is most nearly

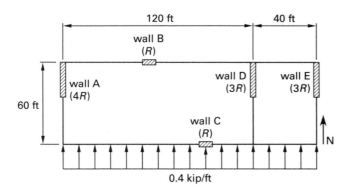

(A) 15 kips

(B) 22 kips

(C) 27 kips

(D) 33 kips

123. The roof framing of a single story commercial building consists of wood joists supported by timber beams and sheathed with a properly nailed and blocked plywood diaphragm. Seismic lateral forces for NS ground motion are shown. Assume sufficiently rigid plywood shear walls 14 ft high and 24 ft long are constructed at lines 1, 2, and 3. Disregard accidental torsion that may be required by code. The axial compression and tension forces in the shear wall boundary members at line 2 under the given loadings are most nearly

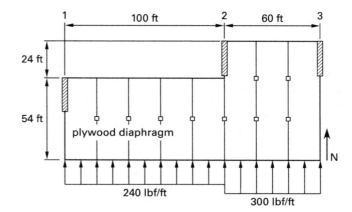

(A) 10 kips

(B) 12 kips

(C) 16 kips

(D) 20 kips

124. For the truss shown, the modulus of elasticity for all members is 29,000 ksi. The cross-sectional area of the members is 8 in². The horizontal deflection at joint D of the truss is most nearly

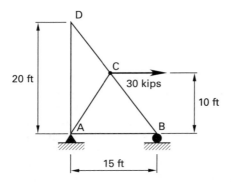

(A) 0.01 in

(B) 0.02 in

(C) 0.04 in

(D) 0.08 in

125. A two-story building is 14 ft from ground to second floor and 12 ft from second floor to roof. The exterior wall projects 3 ft above the roof level to create a parapet. The exterior wall weighs 15 psf, the second floor dead load is 30 psf, and the roof dead load is 20 psf. The building is wood framed with plywood diaphragms and shear walls resisting all lateral forces. The building is situated in seismic performance category D, where the design spectral response acceleration at short periods is 0.6, the design spectral response acceleration at one second period is 0.2, and the importance factor for seismic response is 1.0. Assume the building qualifies as a building frame system with light-frame walls with shear panels. The seismic base shear for north-south (NS) ground motion by the IBC static force procedure, on a working load basis, is most nearly

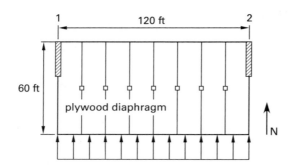

(A) 25 kips

(B) 41 kips

(C) 65 kips

(D) 80 kips

126. The compound beam shown has an internal hinge ($M = 0$) at point B and is simply supported on hinges or rollers at points A, C, and E. The ordinate of the influence line for the bending moment at point D, which is 12 ft to the right of support C, is most nearly

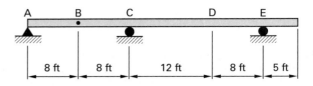

(A) 1 ft-kip/kip

(B) 3 ft-kips/kip

(C) 5 ft-kips/kip

(D) 7 ft-kips/kip

127. A continuous 8 in thick bridge deck is made of reinforced normalweight concrete. It is supported by steel girders spaced 8.5 ft on center, with flange widths of 1 ft. The positive bending moment, per foot of width, for dead weight of the slab is 1.0 ft-kip/ft, and is 0.3 ft-kip/ft for a future wearing course. The deck is continuous over three or more spans and is to be designed by the empirical design approach using the AASHTO *LRFD Bridge Design Specifications* (AASHTO). The factored positive bending moment per foot of deck width that controls deck strength is most nearly

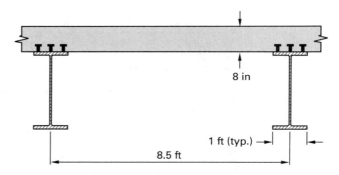

(A) 8 ft-kips/ft

(B) 10 ft-kips/ft

(C) 12 ft-kips/ft

(D) 14 ft-kips/ft

128. The circular shaft shown is subjected to an axial tension force P at its free end and a compressive force of 50 kips at point B. Note that the shaft is hollow between points A and B. The allowable normal tension stress is 22 ksi, the modulus of elasticity is 29,000 ksi, and the maximum allowable elongation is 0.04 in. The maximum allowable value of P is most nearly

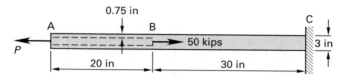

(A) 111 kips

(B) 117 kips

(C) 155 kips

(D) 171 kips

129. A beam is simply supported over a 22 ft span and overhangs the left support 8 ft. Uniformly distributed dead loading of 2 kips/ft and live loading of 3 kips/ft are applied. The live load is positioned to produce maximum shear. The absolute maximum shear force at a point midway between the simple supports (point B′) is most nearly

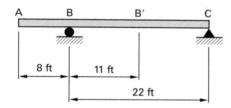

(A) 10 kips

(B) 15 kips

(C) 20 kips

(D) 39 kips

130. In the frame shown, the centroidal area moment of inertia for each leg is 650 in⁴. The modulus of elasticity of steel reinforcement is 29,000 ksi. Neglecting axial and shear deformation, the counterclockwise rotation in radians at joint C is most nearly

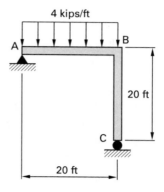

(A) 0.001 rad

(B) 0.004 rad

(C) 0.008 rad

(D) 0.01 rad

131. A normalweight prestressed concrete pile contains four $1/2$ in diameter, 270 ksi, concentrically placed strands that have a 3500 psi compressive stress at release of prestress. Strands are pretensioned to 200 ksi each immediately before being cut. The section is a solid 12 in × 12 in gross cross section. The loss of the prestress due to elastic shortening in this member is most nearly

(A) 5 ksi

(B) 7 ksi

(C) 8 ksi

(D) 9 ksi

132. A pile group consists of 16 piles of the same size and type symmetrically arranged as shown. The group supports a vertical compressive force, P, of 800 kips at an eccentricity of 1.2 ft with respect to the centroid about one principal axis and 1.6 ft about the other principal axis. The maximum axial compression in a pile caused by this loading is most nearly

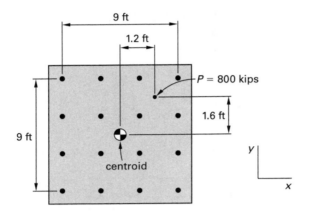

(A) 50 kips

(B) 60 kips

(C) 80 kips

(D) 110 kips

133. The plane truss shown is properly classified as

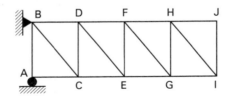

(A) statically determinate and stable

(B) statically determinate and unstable

(C) statically indeterminate to the 1st degree

(D) statically indeterminate to the 2nd degree

134. A thin-walled steel pipe transports water at a pressure of 80 psi. There is negligible restraint to the pipe in its longitudinal direction. If the pipe outside diameter is 60 in and the wall thickness is $^3/_8$ in, the change in pipe diameter caused by the internal pressure is most nearly

(A) 0.005 in

(B) 0.013 in

(C) 0.021 in

(D) 0.025 in

135. The two-span continuous beam shown is subject to a uniformly distributed load of 3.5 kips/ft over both spans. Support B experiences a differential settlement of 0.5 in downward relative to supports A and C. The beam has a moment of inertia of 1630 in^4 and modulus of elasticity of 29,000 ksi in both spans. The reaction at B caused by the uniform load and the differential settlement is most nearly

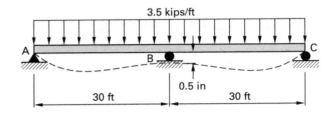

(A) 110 kips

(B) 120 kips

(C) 130 kips

(D) 140 kips

136. The three-span continuous steel beam shown is to be analyzed for the ultimate (i.e., factored) concentrated forces of 25 kips at the midpoint of each span. Neglecting member weight and assuming that strength is controlled by the formation of plastic hinges at critical locations, the required moment capacity is most nearly

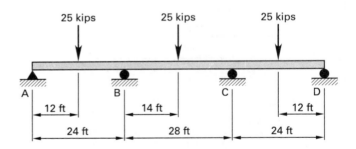

(A) 70 ft-kips

(B) 80 ft-kips

(C) 90 ft-kips

(D) 100 ft-kips

137. A 0.5 kip weight is dropped from rest from 4 ft above an initially straight, simply supported beam and contacts the beam at midspan. The beam is a steel section with a moment of inertia of 1600 in^4 about its axis of bending. Assuming linear elastic behavior, neglecting shear deformation, and considering the weight as an equivalent concentrated force at the midspan, the maximum force exerted on the beam is most nearly

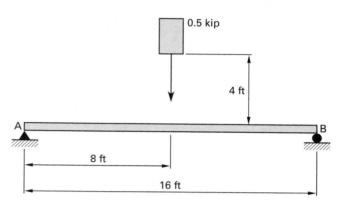

(A) 10 kips

(B) 20 kips

(C) 70 kips

(D) 120 kips

138. An elevated bulk storage bin weighs a total of 30 kips when filled to capacity. The bin can be idealized as a vertical cantilever supported by four steel columns that are fixed at the base and top. Each column has a moment of inertia about its axis of bending of 800 in^4 and is 16 ft long. The modulus of elasticity of steel is 29,000 ksi. Assuming linear elastic behavior, the natural period of vibration for undamped lateral motion of the bin is most nearly

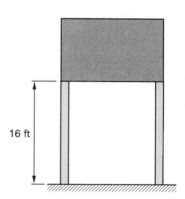

(A) 0.15 sec

(B) 0.3 sec

(C) 0.5 sec

(D) 0.8 sec

139. A steel column section is built by welding two plates 10 in wide by $5/8$ in thick to the flanges of a W12 × 106 to form a doubly symmetrical section. The radius of gyration of the built-up section with respect to its major principal axis is most nearly

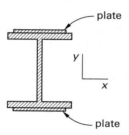

(A) 5.9 in

(B) 6.2 in

(C) 7.1 in

(D) 7.9 in

140. A W24 × 55 has coverplates 8 in wide by $1/2$ in thick symmetrically placed and welded to its top and bottom flanges. The section is subjected to a vertical shear force of 95 kips. Assuming linearly elastic behavior, the horizontal shear flow between the cover plate and flange is most nearly

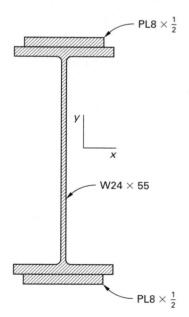

(A) 2 kips/in

(B) 3 kips/in

(C) 5 kips/in

(D) 8 kips/in

141. A reinforced concrete retaining wall is backfilled with a granular material that exerts an active earth pressure equivalent to a 45 pcf fluid. The backfill is level without a surcharge, and the stem extends vertically from the top of the backfill 13 ft to the top of the wall footing. The concrete is normalweight with specified compressive strength of 4000 psi, and grade 60 rebars are specified. The flexural steel percentage is specified to be 0.01. The design is tension-controlled. The required area of reinforcing steel at the bottom of the stem is most nearly

(A) 0.50 in^2/ft

(B) 0.70 in^2/ft

(C) 0.80 in^2/ft

(D) 0.90 in^2/ft

142. For normalweight concrete used in reinforced concrete structures that will be exposed to seawater, the maximum water-to-cementitious-materials ratio, by weight, is required to be

(A) 0.35

(B) 0.40

(C) 0.45

(D) 0.50

143. A road deck is to be supported by a prestressed concrete girder. The girder is 45 ft long, 3.0 ft deep, and 1.5 ft wide. The concrete strength during prestressing is 4 ksi. The local relative humidity is 80%. The girder is 35 days old, and the deck will be placed when the girder is 45 days old. Calculations are being performed to determine the deflection of the girder (including the effects of creep) during the time between the girder fabrication and the deck placement. Most nearly, what is the AASHTO LRFD creep coefficient for the girder?

(A) 0.05

(B) 0.2

(C) 0.5

(D) 0.6

144. A reinforced concrete beam has three No. 9 rebars for both tension and compression as shown. The concrete is normalweight with specified compressive strength of 6000 psi, and rebars are grade 60. Under balanced strain conditions, the total compression force in the three rebars on the compression side is most nearly

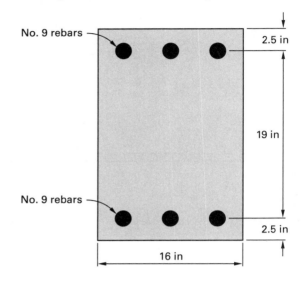

(A) 60 kips

(B) 90 kips

(C) 120 kips

(D) 180 kips

145. A reinforced concrete tied column is subjected to a design axial compression force of 1090 kips that is concentrically applied. Slenderness effects are negligible, and the column is to be designed using ACI 318. Given a specified compressive strength of 5000 psi, grade 60 rebars, and a specified longitudinal steel ratio of 0.02, the smallest square column that will support the load has sides that are most nearly how wide?

(A) 12 in

(B) 16 in

(C) 20 in

(D) 24 in

146. A reinforced concrete tied column is subjected to a design axial compression force of 875 kips and a design bending moment about its strong axis of 175 ft-kips. The column's cross section measures 20 in × 18 in, concrete's specified compressive strength is 4000 psi, steel is grade 60, and the distance from edge of column to center of steel in each face is 3 in. The required area of longitudinal steel is most nearly

(A) 4.0 in²

(B) 8.0 in²

(C) 12 in²

(D) 16 in²

147. A simply supported beam carries a uniform load and is comprised of normalweight concrete with a compressive strength of 3000 lbf/in². Four uncoated No. 9 bars (steel yield strength of 60,000 lbf/in²) provide tension reinforcement. No. 3 stirrups are spaced at 3 in throughout the beam length. The vertical and horizontal clear cover of the stirrups is 1.5 in. The ratio of the required steel area to the provided steel area, $A_{s,\text{req}}/A_{s,\text{prov}}$, is equal to 0.74.

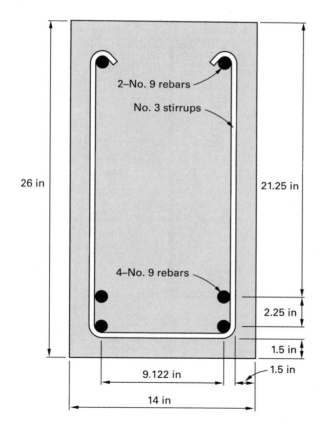

Using applicable length reductions, what is most nearly the minimum required development length permitted by ACI 318 for the bottom tension bars?

(A) 17 in

(B) 42 in

(C) 56 in

(D) 93 in

148. A combined footing constructed of normalweight reinforced concrete is subjected to concentrated forces from column reactions as shown. The columns are 12 in × 12 in, and the footing is 2 ft thick. The maximum bearing pressure beneath the footing, including the footing weight, is most nearly

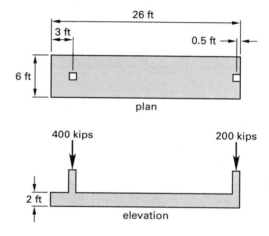

(A) 3 kips/ft²

(B) 4 kips/ft²

(C) 5 kips/ft²

(D) 6 kips/ft²

149. The cantilevered retaining wall shown retains soil with a unit weight of 100 pcf that exerts active earth pressure equivalent to that of a fluid with unit weight 35 pcf. Concrete is normalweight. The factor of safety against overturning is most nearly

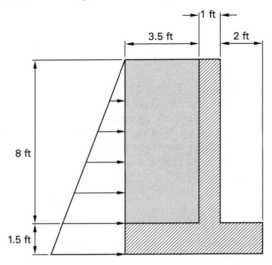

(A) 1

(B) 2

(C) 3

(D) 4

150. A rectangular beam is pretensioned using six $\frac{1}{2}$ in diameter grade 270 strands positioned as shown. The beam is nominally 40 ft long. The concrete is lightweight with a unit weight of 110 pcf and compressive strength of 3500 psi at time of release. The prestress is 200 ksi before release and is 180 ksi immediately after release. The area of pre-stressed reinforcement is $(6)(0.153 \text{ in}^2)$, or 0.918 in^2. The initial midspan upward camber in the beam is most nearly

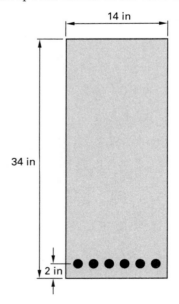

(A) 0.2 in

(B) 0.5 in

(C) 0.7 in

(D) 0.9 in

151. A curved highway ramp has a radius of curvature of 400 ft. The ramp's design speed is 45 mi/hr. The load combination limit state is a strength I load combination. Using only the front axle weight of a standard design truck, what is most nearly the AASHTO LRFD factored centrifugal force for one loaded lane?

(A) 1.2 kips

(B) 2.4 kips

(C) 4.3 kips

(D) 7.6 kips

152. The steel built-up section shown is compact against local buckling in A36 steel. The plastic moment capacity of the section for bending about its strong axis is most nearly

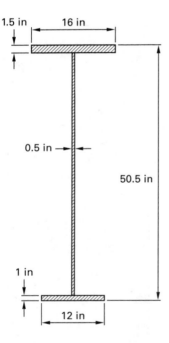

(A) 2200 ft-kips

(B) 2700 ft-kips

(C) 3300 ft-kips

(D) 3900 ft-kips

153. A steel W-section is uniformly loaded to produce bending about its strong axis. The beam supports a uniformly distributed service dead load of 2.5 kips/ft and a service live load of 1.8 kips/ft on a 36 ft simple span. Given that the beam is laterally supported only at the ends and at midspan, the maximum value of the lateral-torsional buckling modificaton factor that may be applied for this beam is most nearly

(A) 1.0

(B) 1.1

(C) 1.2

(D) 1.3

154. Two $L4 \times 4 \times {}^{3}/_{8}$ are welded back to back to a gusset plate of A36 steel using $^{1}/_{4}$ in E70 electrodes. The minimum thickness of the gusset plate that will develop the strength of the two welds is most nearly

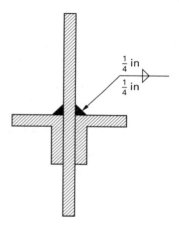

(A) 0.4 in

(B) 0.5 in

(C) 0.6 in

(D) 0.9 in

155. A $W10 \times 49$ member of A992 steel is punched for end connections using $^{3}/_{4}$ in diameter bolts through the flanges and web as shown. Assuming there are at least three bolts in the direction of stress, what is most nearly the allowable axial tensile force permitted by the AISC specification? (ASD options are shown first. LRFD options are given in parentheses.)

(A) 350 kips (510 kips)

(B) 390 kips (580 kips)

(C) 430 kips (650 kips)

(D) 520 kips (690 kips)

156. A built-up compression member consists of two $C12 \times 30$ symmetrically arranged and joined by two $PL^{1}/_{2} \times 11$, one on each side. The plates are welded continuously along their exposed edges to the channels. The plates and channels do not experience differential movement. The member is used as a column, loaded concentrically, with effective lengths of 22 ft about the x-axis and 24 ft about the y-axis. If A36 steel is used, what is most nearly the design load for the column? (ASD options are shown first. LRFD options are given in parentheses.)

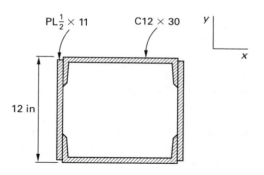

(A) 530 kips (790 kips)

(B) 600 kips (850 kips)

(C) 650 kips (900 kips)

(D) 700 kips (950 kips)

157. A 24F glulam timber beam (with adjusted allowable bending stress of 2400 psi) spans 32 ft and is simply supported. The beam is braced laterally along its compression flange for half its length, but is not braced laterally over the other half, except at the support at the end of the beam. The beam is constructed using visually graded lumber and is $5^1/8$ in wide × 22.5 in deep, which are the actual, not nominal, dimensions. It is subjected to loads of normal duration (i.e., 10 yr cumulative duration) and environmental conditions satisfying dry use. The elastic modulus with respect to the strong axis is 1.6×10^6 psi, and for beam stability calculations is 0.83×10^6 psi. The maximum bending moment due to a uniformly distributed load that can be resisted by this member, based on the NDS, is most nearly

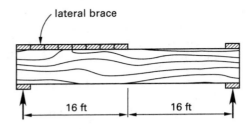

(A) 60 ft-kips

(B) 70 ft-kips

(C) 80 ft-kips

(D) 90 ft-kips

158. A diaphragm is constructed using Structural I grade plywood nailed to 2 in nominal wide framing in the pattern shown. The diaphragm must transfer lateral service wind of 320 lbf/ft to three shear walls. The plywood thickness and nail pattern are adequate to transfer the maximum shear as a blocked diaphragm, and the plywood can transfer 240 lbf/ft as an unblocked diaphragm (with fasteners at their maximum spacing). The largest region between walls B and C (as measured from either side) over which an unblocked diaphragm is adequate is most nearly

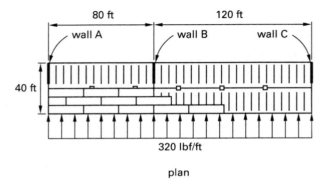

plan

(A) 10 ft to 110 ft

(B) 20 ft to 100 ft

(C) 30 ft to 90 ft

(D) 50 ft to 70 ft

159. A W21 × 93 steel girder supporting a traveling crane is strengthened by welding $3/4$ in coverplates to top and bottom flanges using $1/4$ in E70 fillet welds. The welds are continuous and include transverse welds at the termination points, which occur 3 ft from each end of the girder. The girder is expected to experience approximately 50 applications of maximum stress daily during its design life of 25 yr. For these conditions, the stress range in the tension flange adjacent to the termination point is limited by the AISC specification to

(A) 9.5 ksi

(B) 12 ksi

(C) 19 ksi

(D) 21 ksi

160. A reinforced concrete masonry wall spans vertically 15 ft 4 in from the first floor to the roof. The wall is non-load-bearing and is governed by a wind pressure of 20 psf acting inward or outward. The wall's specified compressive strength is 1500 psi with special inspection, nominal 8 in masonry units grouted solid, grade 60 rebars, and bars centered in the wall. The steel's internal lever arm resisting flexure is assumed to be $7/8$ of the steel's effective depth, d. The required area of flexural reinforcement per foot of wall by IBC allowable stress design is most nearly

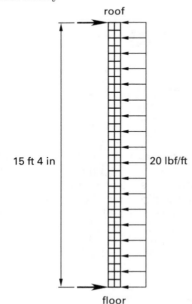

(A) 0.04 in^2/ft

(B) 0.09 in^2/ft

(C) 0.12 in^2/ft

(D) 0.22 in^2/ft

Afternoon Session Transportation

161. The circular curve shown represents the centerline of a two-lane rural highway that passes through level surroundings. The highway has 12 ft lanes and 6 ft shoulders. The arc basis degree of curvature is 2°, and the design speed is 40 mph. The point of intersection (PI) is located at 423,968.68 N, 268,236.42 E. What are most nearly the coordinates of the point of tangent (PT)?

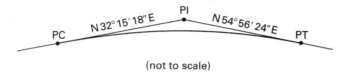

(not to scale)

(A) 424,297.26 N, 268,715.97 E

(B) 424,298.78 N, 268,706.80 E

(C) 424,309.47 N, 268,707.33 E

(D) 424,897.66 N, 268,136.44 E

162. The maximum service flow rate of a four-lane freeway is 1350 pcphpl, and the free-flow speed is 65 mph. The level of service (LOS) at which the freeway operates is most nearly

(A) LOS B

(B) LOS C

(C) LOS D

(D) LOS E

163. The relationship between the average travel speed, S, in mi/hr and the density, D, in veh/mi for an urban road is given by the following relationship.

$$S = 65 \ \frac{\text{mi}}{\text{hr}} - \left(0.42 \ \frac{\frac{\text{mi}}{\text{hr}}}{\frac{\text{veh}}{\text{mi}}} \right) D$$

Assume undersaturated traffic conditions. When the average travel speed is 50 mph, the flow rate is most nearly

(A) 160 vph

(B) 1800 vph

(C) 2300 vph

(D) 3200 vph

164. A circular curve is designed as part of a two-lane rural highway that passes through level surroundings. The highway will have 12 ft lanes with 6 ft shoulders. The design speed is 40 mph and the curve degree of curvature is 2°. A reasonable recommendation of the resulting rate of superelevation is most nearly

(A) 0.02 ft/ft

(B) 0.03 ft/ft

(C) 0.04 ft/ft

(D) 0.05 ft/ft

165. The centerline of a circular curve in a two-lane roadway is shown. Each lane is 12 ft wide. There is no shoulder. The PC station is sta 12+40. The curve radius is 2080 ft. The deflection angle is 60°. What is most nearly the PT station?

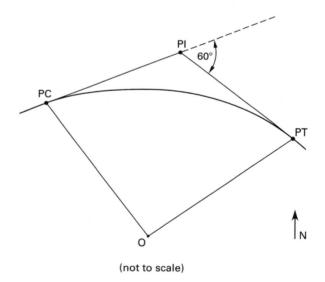

(not to scale)

(A) sta 22+18

(B) sta 24+38

(C) sta 34+18

(D) sta 48+18

166. A rural two-lane highway passes through level terrain. The highway has 12 ft lanes with 6 ft shoulders. The design speed is 40 mph. The approximate minimum required stopping sight distance is most nearly

(A) 92 ft

(B) 160 ft

(C) 300 ft

(D) 450 ft

167. A two-lane roadway is to be superelevated around a circular curve as shown. The axis of rotation will be the centerline. The criteria for the superelevation rate and runoff length are given in the following table. Tangent runouts are twice the runoff lengths listed. Each lane is 12 ft wide, and there is no shoulder. The PC station is sta 12+40. The curve radius is 2080 ft, and the deflection angle is 60°. What is most nearly the station of the beginning of superelevation transition?

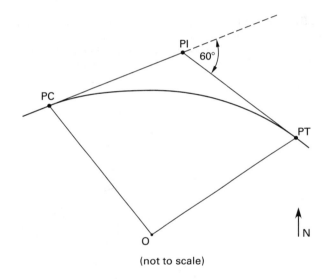

(not to scale)

curve radius (ft)	superelevation (ft/ft)	runoff in the curve (ft)	
		2 lanes	4 lanes
22,920	NC	0	0
11,460	NC	0	0
7640	RC	0	0
5730	0.020	150	150
3820	0.028	150	150
2865	0.035	150	150
2290	0.040	150	150
1910	0.045	150	160
1640	0.048	150	170
1430	0.052	150	180
1145	0.056	150	200
955	0.059	150	210

(A) sta 6+40

(B) sta 9+40

(C) sta 11+40

(D) sta 11+90

168. A two-lane roadway includes a circular curve as shown. Each lane is 12 ft wide, and there is no shoulder. The curve length is 2000 ft, and the PC station is sta 13+50. The roadway is superelevated around the curve. The axis of rotation is at the centerline. The rate of superelevation is 0.045 ft/ft, and the runoff length in the curve is 150 ft. The roadway profile is on a constant uphill grade of 0.750%. The elevation of the centerline at the PC is 170 ft. What is most nearly the elevation of the outside (higher) edge of the roadway at the midpoint along the curve?

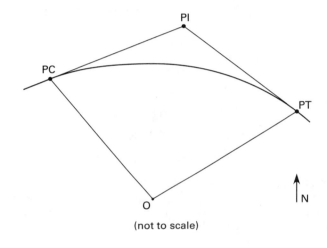

(not to scale)

(A) 170 ft

(B) 176 ft

(C) 177 ft

(D) 178 ft

169. The following traffic volume counts were taken on a road during the peak hour.

time period	volume (veh)
4:30–4:45 p.m.	195
4:46–5:00 p.m.	163
5:01–5:15 p.m.	157
5:16–5:30 p.m.	178

The peak hour factor (PHF) is most nearly

(A) 0.28

(B) 0.89

(C) 0.97

(D) 1.0

170. A six-lane rural freeway has an ideal free-flow speed of 70 mph. Each lane is 12 ft wide. A few sections have grades greater than 3%, but none of these sections are longer than $^1/_4$ mi. The directional weekday volume is 2500 vph in one direction and is mostly commuter traffic. The traffic stream consists of 14% trucks, 8% buses, and 4% recreational vehicles. The peak hour factor from previous volume studies is 0.85. The 15 min passenger car equivalent flow rate, under the prevailing conditions described, is most nearly

(A) 1100 pcphpl

(B) 1110 pcphpl

(C) 1170 pcphpl

(D) 1190 pcphpl

171. An eight-lane freeway near an urban area has an ideal 65 mph free-flow speed. There is one interchange approximately every 2 mi. Each lane is 11 ft wide. The minimum clear distance between overpass abutments, curve rails, and other roadside obstructions is 2 ft on both shoulders. A recent traffic study estimated the 15 min passenger car equivalent flow rate to be 1938 pcphpl. What is most nearly the level of service (LOS) at which the freeway is operating?

(A) LOS B

(B) LOS C

(C) LOS D

(D) LOS E

172. A new freeway with an ideal 70 mph free-flow speed is being designed. The freeway will pass through level terrain in suburban areas. The interchange frequency will be approximately one per mile. A traffic study estimates the 15 min passenger car equivalent flow rate to be 1910 pcphpl based on a six-lane freeway. Base conditions apply. What is most nearly the minimum number of lanes in each direction needed to provide LOS C?

(A) 2

(B) 3

(C) 4

(D) 5

173. A six-lane freeway with an ideal 60 mph free-flow speed passes through rural areas. A traffic study estimates the current 15 min passenger car equivalent flow rate to be 1900 pcphpl. The flow rate is expected to grow at a rate of 5% per year. What is most nearly the number of years before the freeway starts operating at capacity?

(A) 1 yr

(B) 2 yr

(C) 3 yr

(D) 4 yr

174. The approach tangent to an equal-tangent vertical curve has a slope of +3%. The slope of the departure tangent is −2%. These two tangents intersect at sta 26+00 and elevation 231.00 ft. A set of subway rails passes below and perpendicular to the curve at sta 28+50 on the departure side. The maximum elevation of the railbed at that point is 195.00 ft. The length of the vertical curve is 16.48 sta. What is most nearly the minimum distance between the railbed and the roadway surface?

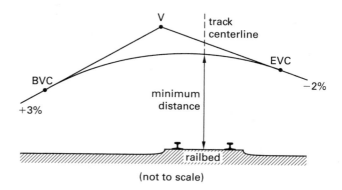

(not to scale)

(A) 5.0 ft

(B) 17 ft

(C) 26 ft

(D) 33 ft

175. A rural collector highway includes a crest vertical curve designed to connect a 2% grade with a −6% grade. The two vertical tangents intersect at sta 32+40.52 and elevation 456.61 ft. The design speed is 55 mph, and the actual traffic speeds are expected to be close to the design speed. What is most nearly the minimum length of the vertical curve for minimum stopping sight distance?

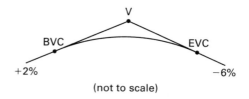

(not to scale)

(A) 300 ft

(B) 500 ft

(C) 700 ft

(D) 900 ft

176. A vertical curve connects two tangents. The approach tangent has a slope of +3%. The slope of the departure tangent is −2%. These two tangents intersect at sta 26+00 and elevation 231.00 ft. If the length of the vertical curve is 16.48 sta, what is most nearly the elevation of the EVC?

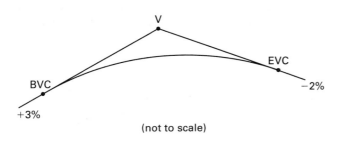

(not to scale)

(A) 100 ft

(B) 200 ft

(C) 210 ft

(D) 250 ft

177. A four-lane highway runs north-south and passes through suburban areas. The preliminary design of the north-bound lanes includes a vertical curve with a length of 22.00 sta. The curve connects a +3% grade with a −5% grade. The two vertical tangents intersect at sta 91+70 and elevation 1453.61 ft. The design speed is 65 mph. What is most nearly the station of the highest point on the curve (i.e., at the turning point)?

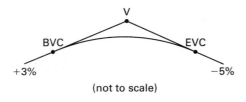

(not to scale)

(A) sta 77+95

(B) sta 88+95

(C) sta 94+45

(D) sta 99+95

178. A two-lane highway is planned for a rural area. The horizontal layout of the highway includes four simple curves. The design criteria for the simple curves include a design speed of 60 mph and a rate of super-elevation of 0.1. One of the simple curves shown is 864 ft long with a radius of 1100 ft. The simple curve connects two tangents that deflect at an angle of 45° and intersect at sta 1500+00. What is most nearly the distance from the point of intersection of the two tangents (vertex) to the midpoint of the curve?

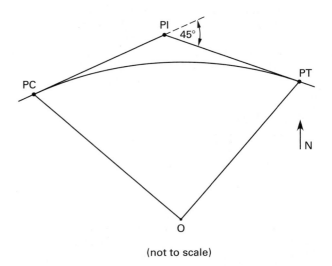

(not to scale)

(A) 82 ft

(B) 84 ft

(C) 91 ft

(D) 460 ft

179. The vertical alignment of a four-lane rural highway includes the vertical curve shown, which connects a −4% grade and a +1% grade. The point of vertical intersection of the grades occurs at sta 28+30 and has an elevation of 2231.31 ft. The design speed is 70 mph, and the length of the curve is 900 ft. What is most nearly the elevation of the lowest point on the curve (the turning point)?

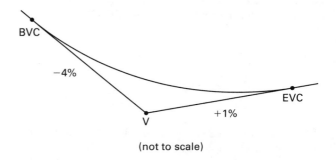

(not to scale)

(A) 2221 ft

(B) 2231 ft

(C) 2233 ft

(D) 2235 ft

180. A 7 mi rural two-lane highway has a directional demand flow rate of 1100 vph. The highway passes through rolling terrain. The design includes 11 ft lanes, 4 ft paved shoulders, 40% no-passing zones, and 15 access points. A traffic study has shown that the directional split during peak hours is 60/40, and the peak hour factor is 0.92. The traffic consists of 5% trucks, 3% buses, and 2% recreational vehicles. The base free-flow speed is 60 mph. The highest directional flow rate for the peak 15 min period is most nearly

(A) 720 vph

(B) 790 vph

(C) 1100 vph

(D) 1300 vph

181. Plans outline a new four-lane freeway that will connect two cities through a suburban area. The freeway will serve as an alternative to an existing two-lane minor arterial, which is currently operating at capacity. The freeway will save motorists travel time and will be safer than the existing arterial. The demand function for travel on the new highway is represented by the straight line in the first illustration. The user cost per vehicle, including toll charges and delay costs, is represented by the supply curve in the second illustration. What is most nearly the expected number of vehicles to use the new freeway?

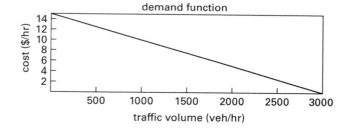

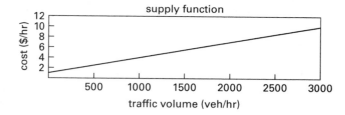

(A) 1000 vph

(B) 1250 vph

(C) 1500 vph

(D) 1750 vph

182. A highway under construction requires fill and cut based on the ground profile. The following mass diagram indicates the net accumulation of cut and fill between two stations. The maximum distance for which there are no additional hauling charges is 500 ft. The unit cost of excavation is $4.35/yd^3, and the overhaul unit cost is $9.75/yd^3 per station. What is most nearly the overhaul cost?

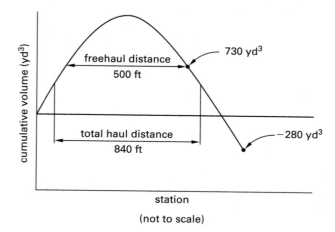

station

(not to scale)

(A) $9300

(B) $11,000

(C) $24,000

(D) $60,000

183. Vehicles arrive at the ticket gate of a parking lot at an average rate of 30 vph. It takes an average of 1.5 min to get a ticket and drive through the gate. The arrival distribution is assumed to be Poisson and the service distribution is assumed to be exponential. There is no space limitation for the vehicles waiting to get a ticket. What is most nearly the number of vehicles expected to be waiting at the gate (i.e., the queue length)?

(A) one vehicle

(B) three vehicles

(C) four vehicles

(D) five vehicles

184. A rural two-lane road has 9 ft lanes and 2 ft unpaved shoulders. A 15 mi section of this road has had 140 traffic accidents (i.e., crashes) over the past 3 yr, 5% of which have been fatal accidents. The current average daily traffic (ADT) is 14,000 vpd. A nearby development is expected to increase the ADT on this section to 18,000 vpd. Construction associated with the development will improve the geometry of this section by adding 2 ft to the existing lanes and 2 ft to the existing shoulders. These improvements are expected to reduce the total and fatal traffic accidents by 25%. What is most nearly the fatal accident rate per 100 million vehicle-miles (HMVM) subsequent to the development and geometric improvements?

(A) 0.6 fatal accidents/HMVM

(B) 1.8 fatal accidents/HMVM

(C) 2.1 fatal accidents/HMVM

(D) 12 fatal accidents/HMVM

185. A traffic study area consists of four zones. The numbers of trip productions and attractions for each zone have been determined in the trip generation process as shown. A calibration process for the gravity model gives calibration values, $F_{i,j}$, which are a function of the travel time between each zone. The calibration values are shown. The socioeconomic conditions are considered to be the same in all zones. What is most nearly the number of trips that are produced in zone 1 and attracted to zone 3, using the gravity model for one iteration?

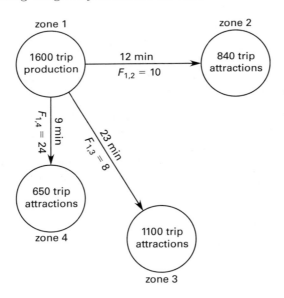

(not to scale)

(A) 300

(B) 430

(C) 740

(D) 1100

186. The following illustration shows a highway network consisting of nodes and links. The average travel times between links are shown. The following table gives the vehicle travel from node 1 to the other nodes based on an origin-destination (O-D) survey. All the links are two-way travel. Use the minimum travel time (all-or-nothing) approach to assign the traffic volumes generated in zone 1 to all available links. What is most nearly the amount of travel in vehicle-minutes for the link between nodes 1 and 5?

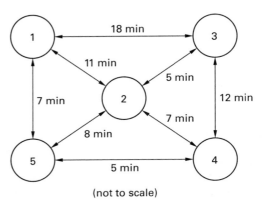

(not to scale)

	traffic volume (veh)
node 1 to node 1	0
node 1 to node 2	100
node 1 to node 3	85
node 1 to node 4	130
node 1 to node 5	115

(A) 800 veh-min

(B) 910 veh-min

(C) 1700 veh-min

(D) 4700 veh-min

187. The travel data in the following table relate to travel between zone 1 and zone 2, using either the auto mode or the transit mode in a suburban area. Use the trip interchange model (Quick Response System (QRS) method) to estimate the mode choice. The exponent value of the model is assumed to be 1.5, and the average income is $30,000.

	auto	transit
distance (mi)	12	10
cost per mile ($)	0.35	0.20
excess time (min)	6	11
speed (mph)	55	45

What is most nearly the expected percent of trips by auto, assuming 120,000 working minutes in a year?

(A) 43%

(B) 57%

(C) 64%

(D) 78%

188. An intersection in a suburban area has had 54 traffic accidents (i.e., crashes) in one year, of which three were fatal and 13 resulted in injuries. The average 24 hr volumes entering each of the four approaches to the intersection are 1250 vpd, 2350 vpd, 730 vpd, and 1920 vpd. What is most nearly the injury accident rate per 10 million vehicles entering the intersection?

(A) 22 accidents/10^7 veh

(B) 57 accidents/10^7 veh

(C) 70 accidents/10^7 veh

(D) 240 accidents/10^7 veh

189. A parking garage is to be located in an urban business area. The garage will be open 6 a.m. to 5 p.m. A total of 400 vehicles will use the garage daily during the hours of operation. 75% of those who use the facility will be commuters with an average parking duration of 8 hr, and the rest are expected to be shoppers with an average parking duration of 3 hr. The parking efficiency, to account for turnovers, is expected to be 0.85. What will be most nearly the number of parking spaces required to meet the parking demand?

(A) 210 spaces

(B) 240 spaces

(C) 290 spaces

(D) 400 spaces

190. A curbed section of a two-lane road passes through an urban area. The section is 1.2 mi long and 40 ft wide curb to curb. Each lane is 11 ft wide. Parking is allowed on both sides of the road. The traffic consists mostly of passenger cars. An on-street parking configuration that minimizes interference with traffic movement is desired. What is most nearly the maximum number of parking spaces on this section of road?

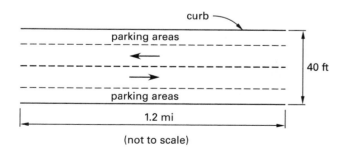

(not to scale)

(A) 290 spaces

(B) 580 spaces

(C) 700 spaces

(D) 740 spaces

191. A weaving section on a highway through an urban area serves the traffic flows shown. Lane widths are 12 ft with no lateral obstructions. The short weaving length, L_S, is 1500 ft. The section is located in level terrain and has an average free-flow speed of 60 mph. The traffic flows represent ideal peak flows and are expressed in passenger cars per hour (pcph). The ideal capacity of the highway section with the same free-flow speed as the weaving section is 1023 pcph per lane. The weaving section is assumed to operate in an unconstrained manner. What is most nearly the average speed of weaving vehicles?

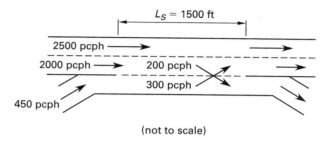

(not to scale)

(A) 38 mph

(B) 44 mph

(C) 50 mph

(D) 68 mph

192. An intersection located in an urban area has a four-phase signal. The four phases are shown. Phases A, B, C, and D have lost times of 4 sec, 4 sec, 2 sec, and 4 sec, respectively. For each phase, the ratios of actual flows to saturation flow, $(v/s)_{ci}$, for all critical lanes, groups, or approaches are also shown. The desired critical ratio of flow to capacity for the whole intersection is 0.90. What is most nearly the signal cycle length using *HCM* procedures?

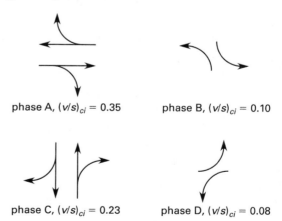

phase A, $(v/s)_{ci} = 0.35$ phase B, $(v/s)_{ci} = 0.10$

phase C, $(v/s)_{ci} = 0.23$ phase D, $(v/s)_{ci} = 0.08$

(A) 50 sec

(B) 80 sec

(C) 90 sec

(D) 120 sec

193. An intersection located in an urban area in level terrain is shown. The maximum allowable (i.e., posted) speed on the approach roads is 45 mph. The width of the intersection is 48 ft for all approaches. The average length of a vehicle is assumed to be 20 ft with a deceleration rate of 11.2 ft/sec². The perception-reaction time of drivers is 2 sec. The area's enforcement philosophy is based on safety enhancement, not revenue generation, and clearance time is included in the all-red interval. What is most nearly the minimum all-red interval for yellow and all-red at this intersection?

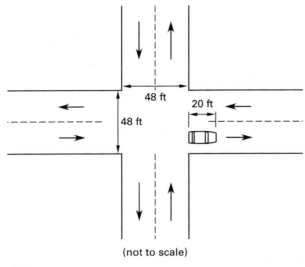

(not to scale)

(A) 2.5 sec

(B) 4.0 sec

(C) 6.0 sec

(D) 8.9 sec

194. Which one of the following statements is NOT true about the geometric design of horizontal curves?

(A) Reverse curves are not recommended because of the possibility of sudden changes to the alignment.

(B) Spiral curves are used between tangents and circular curves or between curves.

(C) Compound curves consist of two or more curves turning in opposite directions.

(D) The most common use of compound curves in highways is for at-grade intersections and ramps of interchanges.

195. Rapid-curing asphalt is formed by cutting asphalt cement with which one of the following distillates?

(A) heavy distillate such as diesel oil

(B) light fuel oil or kerosene

(C) petroleum distillate that easily evaporates, such as gasoline

(D) water

196. Which one of the following statements is NOT true in relation to soils for highway construction?

(A) The main reasons for soil compaction include minimizing future settlement and increasing soil strength.

(B) Sheepsfoot rollers are effective in compacting cohesive soils.

(C) A soil sample with an average CBR value of 48 is considered a superior subgrade to a soil with an average CBR value of 12.

(D) A soil classified A-7-6 (20) is usually rated "good" as a subgrade and is considered suitable as a subbase material.

197. The following illustrations show results obtained using the Marshall mix method in designing an asphalt-concrete mixture. What is most nearly the optimum asphalt content for this mixture before checking design criteria for test limits?

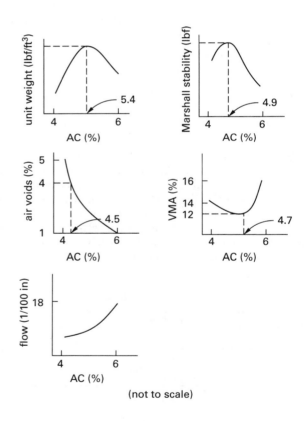

(not to scale)

(A) 4.5%

(B) 4.8%

(C) 5.0%

(D) 5.5%

198. A core is removed from a pavement containing an asphalt-concrete paving mixture designed with 6% asphalt cement. The mass of the core is 1238.5 g in air and 698.3 g submerged. The maximum specific gravity of the mixture is 2.48, and the bulk specific gravity of the combined aggregate in the mixture is 2.64. The percent voids in the mineral aggregate (VMA) is most nearly

(A) 7.7%

(B) 13%

(C) 18%

(D) 77%

199. A road will be constructed in a rural area where the traffic volume is expected to be low and the traffic speed fast. The road is located at latitude 40.9°. The seven day average high air temperature is 31°C, and the one day average low air temperature is −25°C. The standard deviation of both the high and low temperature is ±1.5°C. The pavement surface depth will be 190 mm. Superpave™ procedures will be used to design a suitable asphalt mixture. What will be an appropriate performance grade asphalt binder for this project for 98% reliability?

(A) PG 46-34

(B) PG 52-28

(C) PG 58-16

(D) PG 64-34

200. A flexible pavement for a major arterial road has been designed and applied. The following information is available.

material	thickness (in)	layer coefficient	drainage coefficient
AC surface course	6	0.400	−
untreated granular base	8	0.115	0.50
untreated gravel subbase	10	0.090	0.50

Among the pavement's other characteristics, the equivalent single axle load (ESAL) is 7×10^6; the standard deviation is 0.45; the subgrade resilient modulus is 1050 lbf/in^2; the CBR is 20 for the subbase and 50 for the base; and the elastic modulus of asphalt concrete is 350,000 lbf/in^2. The pavement structure is exposed to moisture levels approaching saturation for 16% of the time, and it takes approximately 70 days for drainage of water. The structural number (SN) of the constructed pavement is most nearly

(A) 2.8

(B) 3.3

(C) 4.2

(D) 4.5

Afternoon Session
Water Resources and Environmental

201. A secondary clarifier is to be designed using the criteria in the following table.

parameter	value
maximum overflow rate	600 gal/day-ft^2
minimum detention time	90 min
maximum weir loading rate	10,000 gal/day-ft
design peak flow	2 MGD
minimum depth	8 ft

What is most nearly the diameter of a single basin that would meet the design criteria?

(A) 45 ft

(B) 55 ft

(C) 65 ft

(D) 75 ft

202. A wastewater treatment plant will use alum to remove 10 mg/L of phosphorus from a flow of 400 L/s. A pilot test determined that 50% above theoretical requirements for alum are needed to effectively remove the phosphorus. Reference data for the alum and phosphorus are given in the following table.

parameter	value
molecular weight of alum	666.7 g/mol
formula for liquid alum	$Al_2(SO_4)_3 \cdot 18H_2O$
alum strength	49%
concentration of alum solution	1.400 kg/L

The volume of alum solution required is most nearly

(A) 6000 L/d

(B) 8000 L/d

(C) 10 000 L/d

(D) 12 000 L/d

203. An activated-sludge wastewater treatment plant receives 400 L/s raw wastewater with 280 mg/L BOD_5 and 220 mg/L total suspended solids (TSS). The final effluent is 20 mg/L BOD_5 and 20 mg/L TSS. The primary clarifier removes 30% of the BOD_5 and 75% of the TSS. The cell yield in the aeration tanks is 60 kg suspended solids produced per 100 kg of BOD_5 removed. No BOD is removed through the secondary clarifier.

The total dry mass of solids produced in the activated sludge process is most nearly

(A) 5000 kg/d

(B) 6000 kg/d

(C) 9000 kg/d

(D) 11 000 kg/d

204. A constructed wetlands design is being reviewed. The design parameters are given in the following table.

parameter	value
length of basin	200 ft
width of basin	2500 ft
fraction of cross section not occupied by plants	0.75
depth of basin	2 ft
average flow rate	40,000 ft^3/day
specific surface area for microbiological activity	4.8 ft^2/ft^3
influent BOD_5	200 mg/L
fraction of BOD_5 not removed by settling at head of system	0.52
rate constant at 20°C	0.006/day
temperature	10°C

The effluent BOD_5 concentration will be most nearly

(A) 48 mg/L

(B) 65 mg/L

(C) 80 mg/L

(D) 95 mg/L

205. The removal of coliform organisms in a small stream is analyzed. The approximate initial die-away rate of the bacteria population is 1500/h. The coefficient of nonuniformity or retardation is 6.15. The stream has a flow of 400 L/s, a depth of 1 m, and a width of 10 m. What is most nearly the percent removal 10 km downstream from the point where the analysis took place?

(A) 30%

(B) 50%

(C) 70%

(D) 90%

206. An anoxic basin will be used to denitrify a wastewater with the characteristics given in the following table.

parameter	value
influent NO_3-N	26 mg/L
effluent NO_3-N	4 mg/L
MLVSS	2500 mg/L
DO	0.2 mg/L
temperature	10°C
specific denitrification rate	0.09 kg NO_3-N/ kg MLVSS·d

The required detention time is most nearly

(A) 5 h

(B) 6 h

(C) 7 h

(D) 8 h

207. A pulp mill discharges a treated effluent to a river where complete mixing occurs quickly below the outfall. The effluent and stream conditions are given in the following table.

parameter	value
effluent flow	0.5 m^3/s
effluent BOD_5 at 20°C	60 mg/L
effluent DO	2 mg/L
effluent temperature	25°C
deoxygenation rate, K_d, base 10, 20°C	0.15 d^{-1}
temperature variation constant θ_d for K_d	1.046
stream flow	6 m^3/s
stream velocity	0.3 m/s
stream BOD_5 at 20°C before mix	4.5 mg/L
stream DO before mix	8.5 mg/L
stream temperature before mix	15°C
reaeration rate, K_r, base 10, 20°C	0.250 d^{-1}
temperature variation constant θ_r for K_r	1.024

The DO of the stream at the critical point is most nearly

(A) 4.8 mg/L

(B) 5.7 mg/L

(C) 6.8 mg/L

(D) 7.6 mg/L

208. A wastewater treatment plant is required to produce an effluent with a coliform count of less than 200/100 mL. Before it is disinfected, the wastewater is found to average 2×10^8/100 mL coliform at a peak hourly flow of 400 L/s. The chlorine contact tank has an effective volume of 1440 m^3. The chlorine residual required to meet the effluent limitation is most nearly

(A) 4.3 mg/L

(B) 5.7 mg/L

(C) 6.5 mg/L

(D) 7.2 mg/L

209. An MPN test gave the results shown in the following table.

serial dilution	sample portion (mL)	no. of positive reactions out of five tubes
0	1.0	5
1	0.1	5
2	0.01	5
3	0.001	2
4	0.0001	1
5	0.00001	0

Using MPN tables, the MPN is

(A) 70/100 mL

(B) 700/100 mL

(C) 7000/100 mL

(D) 70 000/100 mL

210. BOD test results for raw domestic settled wastewater are given in the following table. Each sample is diluted to a total volume of 300 mL.

dilution no.	wastewater volume (mL)	initial DO (mg/L)	5 d DO (mg/L)
1	5	8.0	6.2
2	10	8.2	5.2
3	15	8.4	3.5

For a deoxygenation rate constant (base e) of 0.25/d, the ultimate BOD is most nearly

(A) 120 mg/L

(B) 130 mg/L

(C) 140 mg/L

(D) 150 mg/L

211. A solid waste transfer proposal is being compared to the existing direct haul operation. The characteristics of the two systems are given in the following table.

	value	
parameter	compactor truck direct haul	semi-trailer transfer
capacity in volume	15 yd^3	100 yd^3
transfer station operation cost	–	\$3/yd^3
operation cost	\$40/hr	\$50/hr

The minimum round trip time for which the semi-trailer system would be more economical is most nearly

(A) 85 min

(B) 90 min

(C) 95 min

(D) 100 min

212. A community water supply is contaminated with MTBE and requires treatment to protect residents. The reference dose oral route for MTBE is 0.005 mg/kg·d. An adult has a body mass of 70 kg and an average daily intake of 2 L. For 120 d/yr exposure and complete absorption, the maximum water concentration to protect an adult is most nearly

(A) 0.03 mg/L

(B) 0.36 mg/L

(C) 0.53 mg/L

(D) 1.3 mg/L

213. Pumping tests for a homogeneous aquifer are conducted as shown in the following table. Assume the drawdown compared to aquifer thickness is small, and the length of time of the pumping test is relatively long.

parameter	value
duration of pumping test	3 d
pumping rate	50 L/s
well diameter	450 mm
radius of influence	300 m
elevation of top of aquifer	175 m
elevation of bottom of aquifer	55 m
water surface elevation at 300 m from well	100 m
drawdown in well	3.5 m
elevation of well screen	55–100 m

The coefficient of permeability of the aquifer is most nearly

(A) 200×10^{-6} m/s

(B) 400×10^{-6} m/s

(C) 600×10^{-6} m/s

(D) 800×10^{-6} m/s

214. A groundwater has the chemical analysis given in the following table.

constituent	concentration
Ca^{+2}	90 mg/L
Mg^{+2}	38 mg/L
Na^+	9.7 mg/L
K^+	5.2 mg/L
Fe^{+2}	0.10 mg/L
HCO_3^-	383 mg/L
SO_4^{-2}	39 mg/L
Cl^-	36 mg/L
F^-	0.2 mg/L
NO_3^-	2.2 mg/L
SiO_2	16 mg/L
total dissolved solids	432 mg/L
pH	7.0

The total hardness as $CaCO_3$ of the water is most nearly

(A) 380 mg/L

(B) 420 mg/L

(C) 460 mg/L

(D) 500 mg/L

215. A soil of great depth and approximately uniform permeability is to be drained with drains uniformly spaced with inverts at 3.5 m below the water table. The water level in the drains will be at the crown of the pipe. The average permeability of the sandy soil is 350 m/yr, and groundwater will flow radially to the drains. For a drain length of 1200 m and a drain spacing of 100 m, the total flow from each drain is most nearly

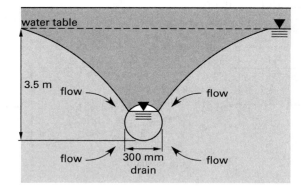

(A) 0.015 m^3/s

(B) 0.020 m^3/s

(C) 0.025 m^3/s

(D) 0.030 m^3/s

216. The following table provides the test results for the discharge coefficients for sections of a broad-crested spillway.

ratio of design head to test head, H/H_o	ratio of design discharge coefficient to test discharge coefficient, C/C_o
0.2	0.84
0.4	0.90
0.6	0.93
0.8	0.97
1.0	1.00
1.2	1.04
1.4	1.04
1.6	1.09
1.8	1.11

The test discharge coefficient for a broad-crested design is 2.20 $\text{ft}^{1/2}/\text{sec}$ for a head of 10 ft. The discharge for a design head of 16 ft and a length of 16 ft is most nearly

(A) 2000 ft^3/sec

(B) 2500 ft^3/sec

(C) 3000 ft^3/sec

(D) 3500 ft^3/sec

217. An outlet pipe from a reservoir (square edge inlet) consists of two sections of concrete pipe as shown. The discharge through a rotary valve (head loss coefficient of 10) at point 3 is to the atmosphere. The water temperature is 50°F. The water's kinematic viscosity is $1.410 \times 10^{-5} \text{ ft}^2/\text{sec}$. What is most nearly the discharge through the valve?

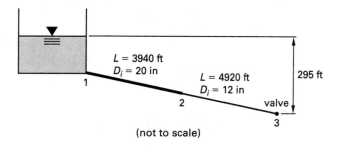

(not to scale)

(A) 4.0 ft^3/sec

(B) 9.0 ft^3/sec

(C) 13 ft^3/sec

(D) 18 ft^3/sec

218. A newly developed 150 ac single-family residential subdivision is drained by an intermittent, straight stream with a uniform cross section. The area in flow is approximately rectangular, 4 ft wide, and 6 ft deep. The streambed is overgrown with weeds and has projecting stones. The elevation change is measured as 1.54 ft over a total length of 21+35 sta. The capacity of the stream when flowing full is most nearly

(A) 36 ft^3/sec

(B) 59 ft^3/sec

(C) 73 ft^3/sec

(D) 120 ft^3/sec

219. A pump station serving a small subdivision will lift water from a clear well of varying water surface elevation. The pump will start at elevation 100 ft and shut off at elevation 120 ft. The elevation of the pump centerline is 90 ft, and the elevation of the discharge reservoir is maintained at a constant 160 ft. The equivalent length of the discharge pipe is 2000 ft of 12 in diameter pipe, and the equivalent length of the suction pipe is 1500 ft of 16 in diameter pipe. Suction and discharge pipes are cement-lined cast iron. The pump performance data are given in the following table.

flow rate (gpm)	total dynamic head (ft)
500	96
1000	88
1500	76
2000	60
2500	36

The pump capacity range for the given conditions will be most nearly

(A) 600-1250 gpm

(B) 1100-1850 gpm

(C) 1350-1700 gpm

(D) 1650-2000 gpm

220. For the condition at which there is no flow into or out of the reservoir at node 2, which of the following statements are true for the pipe network system shown in the illustration?

I. The pressure head at node 6 will be the water surface elevation at the reservoir (100 ft) minus the pipe friction loss in pipes 5 and 6.

II. The flow in pipe 2 will always be $\frac{2}{3}Q_1$.

III. The pressure head at node 2 will be 100 ft plus the pipe friction loss in pipe 1.

IV. The total head loss in pipes 1 and 2 must equal the total head loss in pipes 3 and 4.

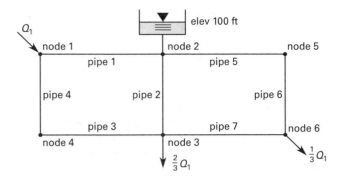

(A) I, II, and IV

(B) I and III

(C) I and IV

(D) II, III, and IV

221. A control structure on the outlet of a storm water detention pond contains a 1 ft diameter pipe set below a rectangular spillway as shown. The spillway has a coefficient of 0.62, and end contractions are suppressed. The pipe protrudes 2 ft into the retention pond. Assume a coefficient of discharge of 0.72. The water surface elevation, h, above the pipe centerline at which the spillway and the pipe will discharge at equal rates is most nearly

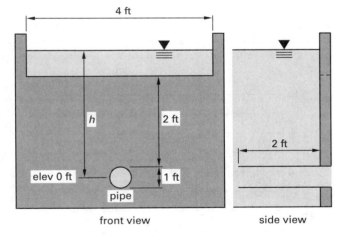

front view   side view

(not to scale)

(A) 1.5 ft

(B) 2.6 ft

(C) 3.2 ft

(D) 4.5 ft

222. A small community is being analyzed to determine its Insurance Services Office (ISO) fire insurance rating. The domestic and commercial water demands and the fire rating demand characteristics of the community are given in the following list. The ISO occupancy combustible factor, O, exposure factor, X, and communication factor, P, are 1.0, 0, and 0, respectively, for all zones.

design population	9000
average annual demand	180 gpcd
large industrial demand	80 gpm in each zone

	zone 1	zone 2	zone 3
average building separation, two-story single family	25 ft	40 ft	120 ft
essential service structures (hospital, police, emergency services)	20,000 ft^2 class 4	50,500 ft^2 class 5	40,800 ft^2 class 3

The minimum delivery capability that will best enable the community to meet domestic and industrial demands concurrently with the fire insurance rating standards for any zone is most nearly

(A) 2100 gpm

(B) 3100 gpm

(C) 4600 gpm

(D) 5100 gpm

223. A V-channel culvert with an interior angle of 90° provides an open channel for conveyance of flood water under a highway. At a control structure 345 ft downstream, the water depth is 5.2 ft for a flow of 400 ft^3/sec. The channel is concrete lined and has a slope of 0.3%. Flow is not uniform, so it will be necessary to compute the length between 0.2 ft changes in water depth. The water depth at the highway crossing is most nearly

(A) 5.2 ft

(B) 5.5 ft

(C) 6.0 ft

(D) 6.4 ft

224. A pump supplies 10 ft^3/sec of water at 90°F from a pit with a constant water level at elevation 5000 ft. Suction friction loss is 3.6 ft at the operating point. The net positive suction head required (NPSHR) is 10.3 ft. What is most nearly the maximum height the pump centerline can be installed above the pit water level to provide a safety factor of 1.3 against cavitation?

(A) 6.0 ft

(B) 9.0 ft

(C) 12 ft

(D) 15 ft

225. A rectangular open channel has the following characteristics and flow conditions.

width	9 ft
discharge	530 ft³/sec
channel slope	0.004
Manning coefficient	0.01

If a hydraulic jump takes place at the normal depth of flow, the energy loss through the jump is most nearly

(A) 0.09 ft

(B) 0.53 ft

(C) 0.96 ft

(D) 1.8 ft

226. A venturi meter is to be calibrated against a tank of known volume for which a discharge is timed. The tank volume is 6.66 ft³ and is totally emptied in 2.8 sec. The venturi meter characteristics are as follows.

throat diameter	6 in
manometer fluid	mercury
height of manometer fluid	8 in (peak flow)
temperature of water	50°F

If the peak discharge is 150% of the average, the flow coefficient of the meter is most nearly

(A) 0.78

(B) 0.84

(C) 0.91

(D) 0.98

227. A culvert for a highway fill is to carry a peak discharge of 30 ft³/sec, at optimum discharge characteristics, through a 24 in, very smooth concrete pipe with a Manning roughness constant of 0.012. The minimum slope that will achieve this condition is most nearly

(A) 0.008 ft/ft

(B) 0.013 ft/ft

(C) 0.018 ft/ft

(D) 0.020 ft/ft

228. Water is released from a wide sluice gate into a rectangular stilling basin of the same width. The bottom of the basin is horizontal and is constructed of concrete. The average discharge velocity is 62 ft/sec, measured a short distance downstream from the point of minimum discharge depth. The minimum depth is 4.08 ft. Beyond the measurement point, the depth increases gradually. Somewhere downstream, the flow experiences a hydraulic jump to attain the depth of the stilling basin. The depth of the flow just before the hydraulic jump is

4.57 ft. The approximate depth of the stilling basin after the hydraulic jump is most nearly

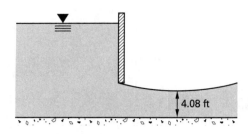

(A) 6.3 ft

(B) 7.9 ft

(C) 25 ft

(D) 27 ft

229. The precipitation at four rain gauge stations in a small watershed is given in the following table.

station	annual precipitation (in)	Storm Alpha precipitation (in)	duration (min)	contributing watershed area (ac)
A	40.5	3.20	57	2000
B	32.8	2.70	57	2500
C	60.7	3.90	57	2800
D	50.3	?	57	2400

The return frequency of storms for this watershed can be found from the following intensity-duration-frequency curves. The return frequency of Storm Alpha is most nearly

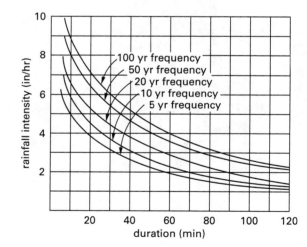

(A) 5 yr

(B) 10 yr

(C) 20 yr

(D) 50 yr

230. A small watershed is covered predominately in alfalfa during the month of May. The watershed has a crop coefficient, k_c, of 0.35 and a soil coefficient, k_s, of 0.80 for a soil with good watering characteristics. The mean wind run in May is 260 km/d, and the mean temperature is 20°C. The mean relative humidity for May is 30% for the watershed. The net radiation for May is 250 W/m². The weighting factors for radiation and vapor transport are 0.7 and 0.3, respectively. The evapotranspiration for the alfalfa watershed is most nearly

(A) 1 mm/d

(B) 3 mm/d

(C) 6 mm/d

(D) 8 mm/d

231. Maximum total precipitation data for a small watershed are given in the following table for 5 min and 20 min subdurations and 10 yr of record.

5 min subduration (cm)	20 min subduration (cm)
1.8	2.1
1.2	2.4
1.0	2.3
1.9	3.1
0.8	2.9
0.7	2.5
1.3	2.8
1.5	2.6
1.2	3.2
0.9	2.4

For a 15 min duration storm, the percent increase in intensity from a 2-yr to a 10-yr storm is most nearly

(A) 10%

(B) 20%

(C) 30%

(D) 40%

232. Sheet flow occurs across a large, impervious parking surface that is smooth and uniform. The surface is 150 ft along the direction of a 6% slope. For a temperature of 60°F and a rainfall intensity of 1.3 in/hr, the velocity of flow at the end of the parking surface is most nearly

(A) 0.20 ft/sec

(B) 0.80 ft/sec

(C) 1.3 ft/sec

(D) 2.2 ft/sec

233. A natural channel and flood plain are shown in the following illustration. During a flood, the river overflows its banks into the east and west portions of the natural flood plain. The river channel is in good condition, but the west flood plain is filled with stones and weeds, and the east flood plain is very poor. The channel and flood plain slopes are approximately 0.002 ft/ft. During a steady flow period when the water depth can be measured as shown, the total flood discharge is most nearly

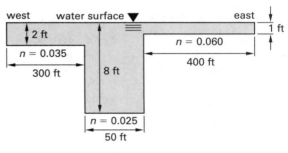

(not to scale)

(A) 4000 ft³/sec

(B) 5000 ft³/sec

(C) 6000 ft³/sec

(D) 7000 ft³/sec

234. The size of a raw water storage reservoir is designed to meet a 2 yr drought condition with negligible precipitation. The stream flow available for use by the treatment plant was analyzed for the low flow conditions as shown in the following table for selected subdurations. Also shown are the evaporation/seepage losses and the design demands for the corresponding subdurations. The area of the reservoir after construction will be 4 km².

subduration (d)	average inflow rate (Mm³/d)	evaporation/ seepage rate (mm)	design demand rate (Mm³/d)
7	0.120	30	0.995
30	0.150	120	0.990
60	0.180	250	0.980
120	0.240	500	0.960
180	0.560	780	0.940
365	1.800	1600	0.900

The required volume of the raw water storage reservoir is most nearly

(A) 22 Mm³

(B) 53 Mm³

(C) 76 Mm³

(D) 88 Mm³

235. A water treatment plant uses alum for coagulation-flocculation at a dose of 12 mg/L. The stoichiometric mass of alum sludge produced is 0.46 kg of sludge per kg of alum dose. The raw water has a turbidity of 5 NTU, and the turbidity after sedimentation is 1 NTU. Jar tests and other studies determined this water has a correlation between turbidity and suspended solids removal, which is not usually the case. The total suspended solids removed is a function of the turbidity removed according to the following formula.

$$\Delta\text{TSS in mg/L} = \Delta\text{NTU}^{1.2}$$

As a coagulant aid, 3 mg/L of clay are added. The removal efficiency of the sedimentation unit is 96% for this range of turbidity. For a flow of 2 m³/s, the mass of solids removed is most nearly

- (A) 1500 kg/d
- (B) 2300 kg/d
- (C) 3200 kg/d
- (D) 4300 kg/d

236. A clarifier for flocculated sedimentation is being designed for a new water treatment plant using a coagulant X on the raw water. Batch settling tests were performed in a 10 ft column. A graph of suspended solids removal (as percent) versus depth and settling times is shown. For a design flow of 1.5 MGD and a scale-up factor of 1.0, what are most nearly the diameter and depth, respectively, of a clarifier that can provide 75% removal?

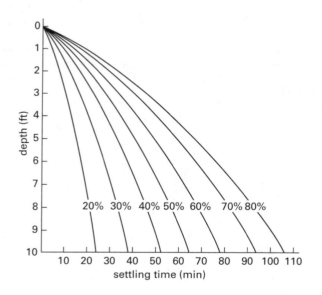

- (A) 32 ft, 6 ft
- (B) 37 ft, 10 ft
- (C) 42 ft, 8 ft
- (D) 45 ft, 12 ft

237. A dual-media filter has the characteristics given in the following table. Assume each layer is uniform.

parameter	property sand	property garnet
depth	280 mm	100 mm
SG	2.65	4.20
average size	1.0 mm	0.40 mm
porosity, ξ	0.50	0.55
shape factor, ϕ	0.8	0.8

For a filtration rate of 4.5 L/s·m² and a temperature of 10°C, the head loss through a clean filter is most nearly

- (A) 0.12 m
- (B) 0.18 m
- (C) 0.26 m
- (D) 0.34 m

238. A clearwell is to be disinfected before being placed in service. The clearwell is 1000 m³ in volume. A 0.5% hypochlorite solution is to be used. The mass of 70% available dry hypochlorite powder required is most nearly

- (A) 4000 kg
- (B) 5500 kg
- (C) 6500 kg
- (D) 7000 kg

239. A contaminant plume of tetrachloroethylene moves toward a river 2 km away. The coefficient of retardation for tetrachloroethylene is given in the table.

travel time (d)	coefficient of retardation
100	0.25
400	0.19
>640	0.17

The depth of the plume is 50 m, and the groundwater elevation is 40 m above the river level. The aquifer is predominately clay, sand, and gravel, with a Darcy coefficient of 0.1 m/d and a porosity of 0.25. What is most nearly the time it will take the contaminant to reach the river?

- (A) 4000 yr
- (B) 6000 yr
- (C) 8000 yr
- (D) 10,000 yr

240. A rapid-mixing basin is designed for a water treatment plant. The preliminary design is given in the table.

parameter	value
basin configuration	square
basin depth	1.5 times width
design flow	3.5 ft³/sec
design velocity gradient	850 ft/sec-ft
design detention time	50 sec
temperature	70°F
impeller power number, laminar flow	65
impeller power number, turbulent flow	5.5
rotational speed	3 rev/sec

The design impeller diameter is most nearly

(A) 1.3 ft

(B) 1.6 ft

(C) 2.0 ft

(D) 2.3 ft

STOP!

DO NOT CONTINUE!

This concludes the Afternoon Session of the examination. If you finish early, check your work and make sure that you have followed all instructions. After checking your answers, you may turn in your examination booklet and answer sheet and leave the examination room. Once you leave, you will not be permitted to return to work or change your answers.

Answer Key

Morning Session

1.	●	B	C	D		11.	A	B	C	●		21.	●	B	C	D		31.	A	B	●	D
2.	A	●	C	D		12.	A	●	C	D		22.	A	B	●	D		32.	A	●	C	D
3.	A	B	C	●		13.	A	B	●	D		23.	A	●	C	D		33.	A	B	C	●
4.	A	B	C	●		14.	A	B	●	D		24.	A	B	●	D		34.	A	B	C	●
5.	A	B	C	●		15.	A	●	C	D		25.	A	B	C	●		35.	A	B	●	D
6.	A	●	C	D		16.	●	B	C	D		26.	A	B	●	D		36.	A	●	C	D
7.	A	B	●	D		17.	●	B	C	D		27.	A	●	C	D		37.	A	B	C	●
8.	A	B	●	D		18.	A	B	●	D		28.	A	B	●	D		38.	A	B	C	●
9.	A	●	C	D		19.	●	B	C	D		29.	A	B	●	D		39.	A	B	●	D
10.	A	B	●	D		20.	●	B	C	D		30.	A	B	●	D		40.	A	B	●	D

Afternoon Session—Construction

41. D	51. A	61. C	71. A
42. B	52. A	62. B	72. B
43. C	53. D	63. D	73. D
44. B	54. C	64. B	74. D
45. B	55. B	65. A	75. C
46. A	56. C	66. B	76. C
47. C	57. A	67. D	77. D
48. D	58. D	68. B	78. B
49. C	59. A	69. C	79. D
50. C	60. D	70. A	80. A

Afternoon Session—Geotechnical

81. B	91. D	101. D	111. B
82. A	92. D	102. C	112. A
83. B	93. B	103. D	113. B
84. B	94. B	104. D	114. A
85. D	95. A	105. C	115. B
86. C	96. C	106. A	116. C
87. C	97. D	107. C	117. A
88. B	98. C	108. B	118. A
89. B	99. C	109. C	119. C
90. C	100. C	110. A	120. B

Afternoon Session—Structural

121. C	131. B	141. D	151. D
122. C	132. D	142. B	152. C
123. B	133. B	143. B	153. D
124. C	134. B	144. D	154. D
125. B	135. C	145. C	155. B
126. C	136. D	146. B	156. A
127. C	137. D	147. B	157. C
128. B	138. A	148. D	158. C
129. B	139. A	149. D	159. A
130. D	140. A	150. B	160. B

Afternoon Session—Transportation

161. B	171. C	181. D	191. C
162. B	172. C	182. C	192. C
163. B	173. D	183. B	193. C
164. B	174. C	184. B	194. C
165. C	175. D	185. B	195. C
166. C	176. C	186. C	196. D
167. A	177. B	187. A	197. C
168. D	178. C	188. B	198. C
169. B	179. D	189. C	199. C
170. B	180. B	190. B	200. B

Afternoon Session—Water Resources and Environmental

201. C	211. A	221. C	231. D
202. B	212. C	222. D	232. B
203. A	213. B	223. C	233. C
204. A	214. A	224. B	234. D
205. D	215. C	225. A	235. B
206. C	216. B	226. A	236. B
207. C	217. B	227. B	237. D
208. D	218. A	228. D	238. D
209. D	219. D	229. D	239. A
210. C	220. C	230. B	240. B

Solutions
Morning Session

1. Calculate the average depth of undercut, D, by summing the undercut depths at each of the corners and dividing the total by the number of corners. Calculate the undercut volume, V, by multiplying the area by the average depth of undercut.

grid number	area (ft^2)	average depth of undercut (ft)	volume (ft^3)
1	2500	2.95	7375
2	1250	2.90	3625
3	2500	2.88	7200
4	2500	2.60	6500
5	2500	2.00	5000
6	2500	1.93	4825
		total	34,525

$$V_{total} = \frac{34{,}525 \text{ ft}^3}{27 \frac{\text{ft}^3}{\text{yd}^3}}$$
$$= 1279 \text{ yd}^3 \quad (1300 \text{ yd}^3)$$

The answer is (A).

2. Flyash is a relatively inexpensive substitute for Portland cement. Its price is about a third of Portland cement's. In concrete, an LOI (loss on ignition) of 2% or less is desired, but an LOI of less than 5% is acceptable. A low loss on ignition means that little organic matter is present in the *pozzolan* (cementitious material).

Class F flyash has a higher LOI and carbon content than class C flyash, and it is usually less than 10% calcium oxide. Class F flyash also absorbs entrained air from the concrete paste and is, therefore, not recommended on important structures, but may be used on interior floor slabs.

Class C flyash has more cement-like compounds, a lower LOI, gains more strength than Class F flyash, and is, therefore, desirable on bridge decks, retaining walls, and other important structures. Class C flyash has 10% to 30% calcium oxide, less than 2% carbon, and an LOI less than 0.5%. Class C flyash is a pozzolan. It can be used in place of cement by as much as 50% by weight.

The answer is (B).

3. Calculate the constituent material weights using the absolute volume (AV) method.

Using the "water required" table given in the problem, a $3/4$ in coarse aggregate (CA) with a 3 in to 4 in slump requires 300 lbf of mix water.

To find the weight of cement, divide the water weight, w, by the cement materials fraction, c.

$$\frac{w}{c} = 0.5$$
$$\frac{300 \text{ lbf}}{c} = 0.5$$
$$c = 600 \text{ lbf}$$

To find the weight of CA, w_{CA}, multiply the dry rodded weight by the volume based on the fineness modulus and the "bulk volume" table given in the problem.

$$w_{CA} = \left(100 \frac{\text{lbf}}{\text{ft}^3}\right)\left(0.60 \frac{\text{ft}^3}{\text{ft}^3}\right)(27 \text{ ft}^3)$$
$$= 1620 \text{ lbf}$$

To find the absolute volume (without fine aggregates, FA), divide the weight of the material by the product of the unit weight of water and the specific gravity of the material.

$$AV_{water} = \frac{300 \text{ lbf}}{\left(62.4 \frac{\text{lbf}}{\text{ft}^3}\right)(1.00)} = 4.81 \text{ ft}^3$$
$$AV_{cement} = \frac{600 \text{ lbf}}{\left(62.4 \frac{\text{lbf}}{\text{ft}^3}\right)(3.15)} = 3.05 \text{ ft}^3$$
$$AV_{CA} = \frac{1620 \text{ lbf}}{\left(62.4 \frac{\text{lbf}}{\text{ft}^3}\right)(2.70)} = 9.62 \text{ ft}^3$$

$$V_{subtotal} = 4.81 \text{ ft}^3 + 3.05 \text{ ft}^3 + 9.62 \text{ ft}^3$$
$$= 17.48 \text{ ft}^3$$

Find the absolute volume of the fine aggregate.

$$AV_{FA} = 27 \text{ ft}^3 - 17.48 \text{ ft}^3 = 9.52 \text{ ft}^3$$

In practice, these volumes would be adjusted according to the workability, durability, and strength properties required for the job.

The answer is (D).

4.

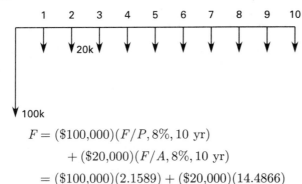

$$F = (\$100{,}000)(F/P, 8\%, 10 \text{ yr})$$
$$+ (\$20{,}000)(F/A, 8\%, 10 \text{ yr})$$
$$= (\$100{,}000)(2.1589) + (\$20{,}000)(14.4866)$$
$$= \$505{,}622 \quad (\$506{,}000)$$

The answer is (D).

5. Float is the amount of time an activity can be delayed without delaying any succeeding activities. It is also known as "slack time."

Total float is defined as the amount of time that an activity in the critical path (e.g., project start) can be delayed without delaying the project completion date.

The answer is (D).

6. The critical path is the longest path through the project activities. For this job, there are four possible scenarios.

1. A-D-G = 3 + 7 + 6 = 16 days
2. B-F = 4 + 9 = 13 days
3. A-E = 3 + 8 = 11 days
4. A-C-F = 3 + 5 + 9 = 17 days

The longest path takes 17 days.

The answer is (B).

7. Anchorage strength in rock is

$$P_u = Ab_r = 2\pi r L b_r$$

P_u is the unfactored anchor capacity, r is the radius of the anchor shaft, L is the length of the anchor shaft, and b_r is the bond strength of rock. This equation represents the shear strength of the limestone over the surface area of the shaft.

$$r = \frac{6 \text{ in}}{2} = 3 \text{ in}$$

$$P_u = 2\pi r L b_r = \frac{2\pi(3 \text{ in})(12 \text{ ft})\left(120 \ \frac{\text{lbf}}{\text{in}^2}\right)\left(12 \ \frac{\text{in}}{\text{ft}}\right)}{1000 \ \frac{\text{lbf}}{\text{kip}}}$$

$$= 325.7 \text{ kips} \quad (320 \text{ kips})$$

The answer is (C).

8. Cycle time refers to the amount of time it takes the backhoe operator to load, dump, and swing back to load another bucket of material.

$$\text{production} = \frac{(\text{capacity})(\text{net hours worked})\left(\frac{\text{sec}}{\text{hr}}\right)}{\text{cycle time}}$$

$$= \frac{\begin{array}{c}(2.0 \text{ yd}^3)(10 \text{ hr} - 1 \text{ hr} - 0.25 \text{ hr} - 0.25 \text{ hr}) \\ \times \left(3600 \ \frac{\text{sec}}{\text{hr}}\right)\end{array}}{40 \text{ sec}}$$

$$= 1530 \text{ yd}^3 \quad (1500 \text{ yd}^3)$$

The answer is (C).

9. Coulomb's equation for active force on soil should be used to solve this problem.

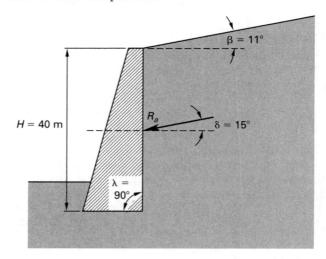

λ is the angle between a horizontal plane and the interior face of the retaining wall. Since the retaining wall surface in contact with the soil is vertical, λ is $90°$.

$$k_a = \frac{\sin^2(\lambda + \phi)}{\sin^2 \lambda \, \sin(\lambda - \delta)\left(1 + \sqrt{\dfrac{\sin(\phi + \delta) \, \sin(\phi - \beta)}{\sin(\lambda - \delta) \, \sin(\lambda + \beta)}}\right)^2}$$

$$= \frac{\sin^2(90° + 30°)}{\sin^2 90° \, \sin(90° - 15°)}$$

$$\times \left(1 + \sqrt{\dfrac{\sin(30° + 15°) \, \sin(30° - 11°)}{\sin(90° - 15°) \, \sin(90° + 11°)}}\right)^2$$

$$= 0.348$$

The active force per unit length of wall is

$$R_a = \tfrac{1}{2}k_a \gamma H^2$$

$$= \left(\frac{1}{2}\right)(0.348)\left(18 \ \frac{\text{kN}}{\text{m}^3}\right)(4.0 \text{ m})^2$$

$$= 50.1 \text{ kN/m} \quad (50 \text{ kN/m})$$

The answer is (B).

10. Before the sand fill is placed on top of the silt deposit, the total stress, water pressure, and effective stress are, respectively,

$$\sigma = \gamma_{\text{sat}} z_A = \left(18 \ \frac{\text{kN}}{\text{m}^3}\right)(11.0 \text{ m})$$

$$= 198 \text{ kN/m}^2$$

$$u = \gamma_w z_w = \left(9.81 \ \frac{\text{kN}}{\text{m}^3}\right)(11.0 \text{ m} - 1.0 \text{ m})$$

$$= 98.1 \text{ kN/m}^2$$

$$\sigma' = \sigma - u = 198 \ \frac{\text{kN}}{\text{m}^2} - 98.1 \ \frac{\text{kN}}{\text{m}^2}$$

$$= 99.9 \text{ kN/m}^2$$

After placing the sand fill on top of the silt deposit, the total stress becomes

$$\sigma_{\text{filled}} = \gamma_{\text{sat}} z_A + \gamma_{\text{sand fill}} z_{\text{sand fill}}$$

$$= \left(18 \ \frac{\text{kN}}{\text{m}^3}\right)(11.0 \text{ m}) + \left(20 \ \frac{\text{kN}}{\text{m}^3}\right)(5.0 \text{ m})$$

$$= 298 \text{ kN/m}^2$$

At the end of the consolidation process, the water pressure is the same as before the placement of the fill.

$$u = 98.1 \text{ kN/m}^2$$

$$\sigma'_{\text{filled}} = \sigma_{\text{filled}} - u = 298 \ \frac{\text{kN}}{\text{m}^2} - 98.1 \ \frac{\text{kN}}{\text{m}^2}$$

$$= 199.9 \text{ kN/m}^2$$

Therefore, the increase in effective stress at point A is

$$\Delta \sigma'_A = \sigma'_{\text{filled}} - \sigma' = 199.9 \ \frac{\text{kN}}{\text{m}^2} - 99.9 \ \frac{\text{kN}}{\text{m}^2}$$

$$= 100 \text{ kN/m}^2$$

The answer is (C).

11. A properly drawn flow net should be similar to the one shown, or at least have approximately the same N_f-to-N_p ratio. (Do not spend a lot of time trying to draw a perfect flow net.)

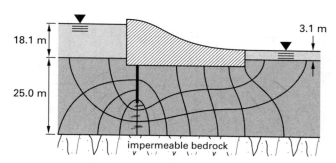

The flow net shows three flow tubes ($N_f = 3$) and 11 head drops ($N_p = 11$). The entire flow is

$$Q = KH\left(\frac{N_f}{N_p}\right)L$$

$$= \left(2 \times 10^{-6} \ \frac{\text{m}}{\text{s}}\right)(18.1 \text{ m} - 3.1 \text{ m})\left(\frac{3}{11}\right)(300 \text{ m})$$

$$= 2.45 \times 10^{-3} \text{ m}^3/\text{s} \quad (2.5 \times 10^{-3} \text{ m}^3/\text{s})$$

The answer is (D).

12. The Boussinesq stress contour chart may be used to solve this problem.

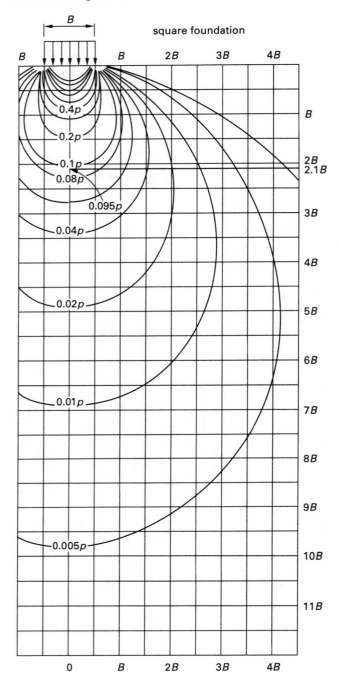

For a column load, N, of 800 kN and a bearing pressure, p, of 400 kN/m^2, the width of the square footing is

$$B = \sqrt{\frac{N}{p}} = \sqrt{\frac{800 \text{ kN}}{400 \ \frac{\text{kN}}{\text{m}^2}}}$$
$$= 1.414 \text{ m}$$

Therefore,

$$\frac{y}{B} = \frac{3 \text{ m}}{1.414 \text{ m}}$$
$$= 2.12$$

The resulting factor from the Boussinesq chart is $0.095p$. The resulting stress is

$$\sigma_y = 0.095p = (0.095)\left(400 \ \frac{\text{kN}}{\text{m}^2}\right)$$
$$= 38 \text{ kN/m}^2$$

For the 100 kN/m^2 foundation bearing pressure,

$$B = \sqrt{\frac{800 \text{ kN}}{100 \ \frac{\text{kN}}{\text{m}^2}}}$$
$$= 2.83 \text{ m}$$

Therefore,

$$\frac{y}{B} = \frac{3 \text{ m}}{2.83 \text{ m}}$$
$$= 1.06$$

The resulting factor from the Boussinesq stress contour chart is $0.3p$. The resulting stress is

$$\sigma_y = 0.3p = (0.3)\left(100 \ \frac{\text{kN}}{\text{m}^2}\right)$$
$$= 30 \text{ kN/m}^2$$

The change in stress at 3 m below the center of the foundation is

$$\Delta\sigma_y = 30 \ \frac{\text{kN}}{\text{m}^2} - 38 \ \frac{\text{kN}}{\text{m}^2}$$
$$= -8 \text{ kN/m}^2 \quad (-10 \text{ kPa})$$

The answer is (B).

13.

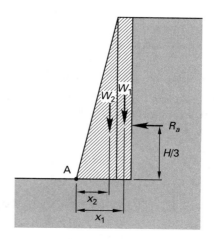

In this case, the only force that tends to resist overturning is the weight of the wall, W. The only force that tends to overturn the wall is the active force, R_a.

The active earth pressure coefficient is

$$k_a = \tan^2\left(45° - \frac{\phi}{2}\right) = \tan^2\left(45° - \frac{31°}{2}\right)$$
$$= 0.32$$

The wall's angle of friction, δ, is $0°$, so the active force can be calculated using the Rankine theory. (The saturated density is used, as this is the maximum density for a drained condition. "Saturated" does not mean "submerged.")

$$R_a = \tfrac{1}{2}k_a\gamma H^2$$
$$= \left(\frac{1}{2}\right)(0.32)\left(19\ \frac{\text{kN}}{\text{m}^3}\right)(5.0\ \text{m})^2$$
$$= 76\ \text{kN/m}$$

The overturning moment per unit length of wall is

$$\sum M_{\text{OT}} = R_a\left(\frac{H}{3}\right) = \left(76\ \frac{\text{kN}}{\text{m}}\right)\left(\frac{5.0\ \text{m}}{3}\right)$$
$$= 126.7\ \text{kN·m/m}$$

The weight of the wall, W, is equal to W_1 plus W_2.

$$W_1 = HD\gamma_{\text{concrete}}$$
$$= (5.0\ \text{m})(1.0\ \text{m})\left(24\ \frac{\text{kN}}{\text{m}^3}\right)$$
$$= 120\ \text{kN/m}$$
$$W_2 = \tfrac{1}{2}HD\gamma_{\text{concrete}}$$
$$= \left(\frac{1}{2}\right)(5.0\ \text{m})(1.5\ \text{m})\left(24\ \frac{\text{kN}}{\text{m}^3}\right)$$
$$= 90\ \text{kN/m}$$

The resisting moment per unit length of wall is

$$\sum M_{\text{resisting}} = W_1 x_1 + W_2 x_2$$
$$= \left(120\ \frac{\text{kN}}{\text{m}}\right)(2.0\ \text{m}) + \left(90\ \frac{\text{kN}}{\text{m}}\right)(1.0\ \text{m})$$
$$= 330\ \text{kN·m/m}$$

The factor of safety against overturning about the toe (point A of the illustration) can be expressed as

$$\text{FS}_{\text{OT}} = \frac{\sum M_{\text{resisting}}}{\sum M_{\text{OT}}} = \frac{330\ \dfrac{\text{kN·m}}{\text{m}}}{126.7\ \dfrac{\text{kN·m}}{\text{m}}}$$
$$= 2.6 \quad (2.5)$$

The answer is (C).

14. Since ϕ is $0°$, the Taylor slope stability chart can be used to solve this problem.

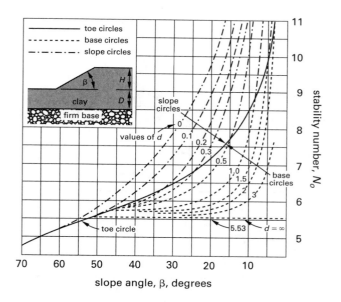

Source: *Soil Mechanics*, NAVFAC Design Manual DM-7.1, 1986, Fig. 2, p. 7.1-319.

The depth factor, d, is the ratio of the vertical distance between the toe of the slope and the firm base below the clay layer, D, to the slope height, H (the depth of the cut).

$$d = \frac{D}{H} = \frac{3.0\ \text{m}}{6.0\ \text{m}}$$
$$= 0.5$$

The Taylor chart shows that, for a depth factor of 0.5 and a slope angle, β, of $55°$, the stability number, N_o, is 5.5.

The cohesive factor of safety for the slope is

$$F_{cohesive} = \frac{N_o c}{\gamma H}$$

$$= \frac{(5.5)\left(50 \ \frac{kN}{m^2}\right)}{\left(19 \ \frac{kN}{m^3}\right)(6.0 \ m)}$$

$$= 2.41 \quad (2.5)$$

The specific weight is not reduced by the specific weight of water because there is no pore pressure (i.e., the soil is above the water table). The soil is not submerged, so the effective specific weight is not used.

The answer is (C).

15. A strength envelope can be constructed using the data given. Mohr's circle can then be constructed tangent to the strength envelope at point 2, as shown. From this arrangement, the major and the minor principal stresses are found to be 95 kPa (say, 100 kPa) and 350 kPa.

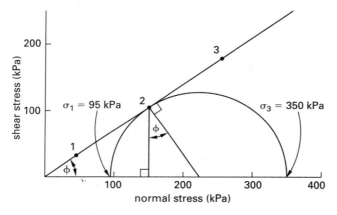

The answer is (B).

16. This problem requires use of the American Association of State Highway and Transportation Officials (AASHTO) classification table for soils and soil-aggregate mixtures.

The percentages of material passing the no. 10, no. 40, and no. 200 sieves are

no. 10 (2.00 mm)	45%
no. 40 (0.425 mm)	35%
no. 200 (0.075 mm)	28%

The liquid limit (LL) and plasticity index (PI) of the material passing the no. 40 sieve are 34 and 13, respectively. Therefore, working from left to right on the AASHTO classification chart, the appropriate subgroup for this soil is A-2-6.

The AASHTO classification also contains a number in parentheses. This number is the group index. For subgroups A-2-6 and A-2-7, the group index is

$$I_g = 0.01(F_{200} - 15)(PI - 10)$$

F_{200} is the percentage of soil passing the no. 200 sieve.

The group index is always reported to the nearest whole number. (If it is calculated to be negative, then it is reported as zero.)

$$I_g = 0.01(F_{200} - 15)(PI - 10)$$

$$= (0.01)(28 - 15)(13 - 10)$$

$$= 0.39 \quad (0, \text{ closest whole number})$$

Therefore, the final AASHTO classification for this soil is A-2-6 (0).

The answer is (A).

17. The resultant is a 32 kips force located $\bar{x}$ from the leftmost load.

$$\bar{x} = \frac{\sum F_i x_i}{\sum F_i}$$

$$= \frac{(8 \ kips)(0 \ ft) + (8 \ kips)(10 \ ft) + (16 \ kips)(22 \ ft)}{8 \ kips + 8 \ kips + 16 \ kips}$$

$$= 13.5 \ ft$$

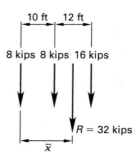

The absolute maximum bending moment occurs under the 16 kips load when the distance from midspan to 16 kips is equidistant from midspan to resultant. Therefore, the distance from midspan to 16 kips is

$$\frac{22 \ ft - 13.5 \ ft}{2} = 4.25 \ ft$$

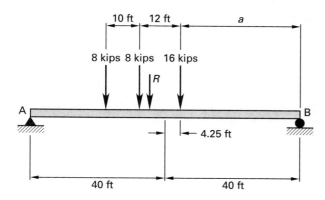

The reaction at the right support is found by summing moments about the left support.

$$\sum M_A = 0 = R_B(80 \text{ ft}) - (32 \text{ kips})(40 \text{ ft} - 4.25 \text{ ft})$$
$$R_B = 14.3 \text{ kips}$$

The bending moment under the 16 kips load is

$$M_{max} = R_B a$$
$$= (14.3 \text{ kips})(40 \text{ ft} - 4.25 \text{ ft})$$
$$= 511 \text{ ft-kips} \quad (510 \text{ ft-kips})$$

The answer is (A).

18.

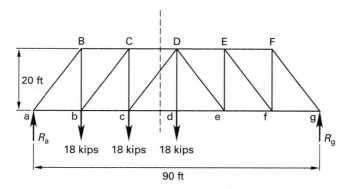

Sum moments about point g. (Assume that the clockwise moment is positive.)

$$\sum M_g = \sum_i r_i F_i = 0$$

$$(90 \text{ ft}) R_a - (18 \text{ kips})(75 \text{ ft} + 60 \text{ ft} + 45 \text{ ft}) = 0$$
$$R_a = 36 \text{ kips}$$

Pass a cutting plane vertically between panel points C and D; consider the left side as the free body. Take moments about point c.

$$\sum M_c = \sum_i r_i F_i = 0$$

$$(20 \text{ ft}) F_{CD} + (36 \text{ kips})(30 \text{ ft}) - (18 \text{ kips})(15 \text{ ft}) = 0$$
$$F_{CD} = -40.5 \text{ kips} \quad (-40 \text{ kips})$$

The answer is (C).

19. Since the resultant compression force acts within the middle one-third of the cross section, the gross cross-sectional area is effective in resisting the combined axial and bending stresses.

$$A = bh = (32 \text{ in})(32 \text{ in})$$
$$= 1024 \text{ in}^2$$
$$I = \frac{bh^3}{12} = \frac{(32 \text{ in})(32 \text{ in})^3}{12}$$
$$= 87{,}381 \text{ in}^4$$

P is the axial compression force, e is the eccentricity, and c is the distance from the center of the column to the edge.

$$f_c = \frac{-P}{A} - \frac{Pec}{I}$$
$$= \frac{-110 \text{ kips}}{1024 \text{ in}^2} - \frac{(110 \text{ kips})(5 \text{ in})(16 \text{ in})}{87{,}381 \text{ in}^4}$$
$$= -0.208 \text{ kips/in}^2 \quad (0.2 \text{ kips/in}^2)$$

The answer is (A).

20. The area of one no. 10 bar is 1.27 in^2. The yield strength of grade 60 rebar is $60{,}000 \text{ lbf/in}^2$.

The height of the compressive stress block is

$$a = \frac{A_s f_y}{0.85 f'_c b} = \frac{(4)(1.27 \text{ in}^2)\left(60{,}000 \ \frac{\text{lbf}}{\text{in}^2}\right)}{(0.85)\left(4000 \ \frac{\text{lbf}}{\text{in}^2}\right)(14 \text{ in})}$$
$$= 6.40 \text{ in}$$

The distance from the top surface to the neutral axis is

$$c = \frac{a}{\beta_1} = \frac{6.40 \text{ in}}{0.85} = 7.53 \text{ in}$$

The tension in the steel is

$$\epsilon_t = 0.003\left(\frac{d-c}{c}\right) = (0.003)\left(\frac{18.5 \text{ in} - 7.53 \text{ in}}{7.53 \text{ in}}\right)$$
$$= 0.00437$$

Since $0.002 < \epsilon_t < 0.005$, the section is in the transition region between tension- and compression-control. Since the beam is not spirally wrapped, the strength reduction factor is

$$\phi_{\text{transition}} = 0.48 + 83\epsilon_t = 0.48 + (83)(0.00437) = 0.843$$

The nominal beam strength is

$$
\begin{aligned}
M_n &= A_s f_y \left(d - \frac{a}{2} \right) \\
&= \frac{(4)(1.27 \text{ in}^2) \left(60{,}000 \ \frac{\text{lbf}}{\text{in}^2} \right) \left(18.5 \text{ in} - \frac{6.40 \text{ in}}{2} \right)}{\left(1000 \ \frac{\text{lbf}}{\text{kip}} \right) \left(12 \ \frac{\text{in}}{\text{ft}} \right)} \\
&= 389 \text{ ft-kips}
\end{aligned}
$$

The beam strength is

$$
\begin{aligned}
M_u &< \phi M_n \\
&= (0.843)(389 \text{ ft-kips}) \\
&= 328 \text{ ft-kips} \quad (330 \text{ ft-kips})
\end{aligned}
$$

The answer is (A).

21. Allowing 3 in minimum cover plus 1 in bar diameter for the separation between the no. 8 rebars in each direction, the effective footing depth, d, measured from the footing top surface to the middle of the two criss-crossing bar layers is 16 in.

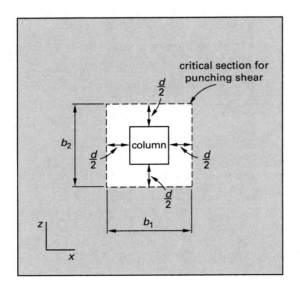

The critical punching shear perimeter is at $d/2$ from the column face. Therefore,

$$
\begin{aligned}
b_o &= 4(h + d) = (4)(14 \text{ in} + 16 \text{ in}) \\
&= 120 \text{ in}
\end{aligned}
$$

For square columns, the controlling punching shear capacity is given in ACI 318.

$$
\begin{aligned}
\phi V_n &= 4\phi\sqrt{f'_c} b_o d \\
&= (4)(0.75)\sqrt{3000 \ \frac{\text{lbf}}{\text{in}^2}} (120 \text{ in})(16 \text{ in}) \\
&= 315{,}488 \text{ lbf} \quad (316 \text{ kips})
\end{aligned}
$$

The design force on the column is uniformly distributed over the 8.5 ft × 8.5 ft spread footing. The punching shear excludes the distributed force within the distance $d/2$ to every side of the column face. Therefore,

$$V_u = w_u A_{\text{trib}} \le \phi V_n$$

$$w_u \left((8.5 \text{ ft})(8.5 \text{ ft}) - \left(\frac{14 \text{ in} + (2)\left(\frac{16 \text{ in}}{2} \right)}{12 \ \frac{\text{in}}{\text{ft}}} \right)^2 \right)$$

$$\le 316 \text{ kips}$$

$$w_u = 4.79 \text{ kips/ft}^2$$

$$P_u \le w_u A_f \le \left(4.79 \ \frac{\text{kips}}{\text{ft}^2} \right) (8.5 \text{ ft})(8.5 \text{ ft})$$

$$\le 346 \text{ kips} \quad (350 \text{ kips})$$

The answer is (A).

22. The location of the centroid of the fastener group is known from symmetry. The polar moment of inertia of the group is

$$
\begin{aligned}
J &= I_x + I_y \\
&= (4)\left((3 \text{ in})^2 + (6 \text{ in})^2 \right) + (10)\left(\frac{5 \text{ in}}{2} \right)^2 \\
&= 242.5 \text{ in}^2
\end{aligned}
$$

The statically equivalent force system is 36 kips upward through the centroid plus a counterclockwise moment of $(8.5 \text{ in})(36 \text{ kips}) = 306$ in-kips. The critical fastener is an extreme fastener for which the y-component acts down and adds algebraically to the y-component of the direct force. Therefore,

$$R_{py} = \frac{P}{n} = \frac{36 \text{ kips}}{10}$$
$$= 3.6 \text{ kips}$$

$$R_{my} = \frac{Mx}{J} = \frac{(306 \text{ in-kips})(2.5 \text{ in})}{242.5 \text{ in}^2}$$
$$= 3.15 \text{ kips}$$

$$R_{mx} = \frac{My}{J} = \frac{(306 \text{ in-kips})(6 \text{ in})}{242.5 \text{ in}^2}$$
$$= 7.57 \text{ kips}$$

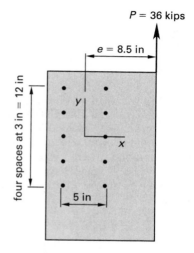

The resultant is found by vector addition of the components.

$$R = \sqrt{R_{mx}^2 + (R_{my} + R_{py})^2}$$
$$= \sqrt{(7.57 \text{ kips})^2 + (3.15 \text{ kips} + 3.6 \text{ kips})^2}$$
$$= 10.1 \text{ kips} \quad (10 \text{ kips})$$

The answer is (C).

23. The flexible diaphragm behaves as a simple beam between the supporting shear walls. The maximum bending moment occurs in the 80 ft span between shear walls on lines B and C.

$$M_{max} = \frac{wL^2}{8} = \frac{\left(320 \dfrac{\text{lbf}}{\text{ft}}\right)(80 \text{ ft})^2}{8}$$
$$= 256{,}000 \text{ ft-lbf}$$

The moment is resisted by the chords, which are nominally 55 ft apart. Therefore,

$$C = T = \frac{M_{max}}{b}$$
$$= \frac{256{,}000 \text{ ft-lbf}}{55 \text{ ft}}$$
$$= 4654 \text{ lbf} \quad (4700 \text{ lbf})$$

The answer is (B).

24. The Mohr's circle drawn for the state of stress on any two orthogonal faces plots as a point at $(\sigma, 0)$. Thus, the radii of the stress circles are all zero, which means that the shear stress is zero for all possible orientations of the element. So, there can be no shear slip in the element. Plastic deformation can occur in steel only in the presence of shear stresses; therefore, ductile behavior cannot occur. Since the shear stress is zero under this triaxial state of stress, failure can occur only by cleavage.

The answer is (C).

25. The bearing of the radius, B_r, at the PT is offset $90°$ from the tangent at the PT.

$$B_r = 90° + B_t$$
$$= 90° + \text{N}\,54°\,56'\,24''\,\text{E}$$
$$= \text{N}\,144°\,56'\,24''\,\text{E} \quad (\text{S}\,35°\,3'\,36''\,\text{E})$$

The answer is (D).

26. In a three-leg intersection, there are three conflict points for each of the conflict types (crossing, merging, and diverging).

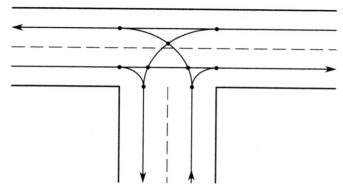

The total number of conflict points is nine.

The answer is (C).

27. The peak hour factor (PHF) is defined as the ratio of the total hourly volume (HV) to the peak rate of flow within the hour (PV). The volume of traffic that passes during the busiest 15 minutes, $V_{15\,min}$, is 750 vph.

$$\begin{aligned} \text{PHF} &= \frac{\text{HV}}{\text{PV}} = \frac{V}{4V_{15\,min}} \\ &= \frac{2400\ \dfrac{\text{veh}}{\text{hr}}}{(4)\left(750\ \dfrac{\text{veh}}{\text{hr}}\right)} \\ &= 0.80 \end{aligned}$$

The answer is (B).

28. Use the Pythagorean theorem and the given coordinates to calculate the required distance.

$$\begin{aligned} D &= \sqrt{\left(N_{cemetery} - N_{curve}\right)^2 + \left(E_{cemetery} - E_{curve}\right)^2} \\ &= \sqrt{\begin{array}{c}(424{,}239.72\ \text{ft} - 424{,}180.59\ \text{ft})^2 \\ + (268{,}498.69\ \text{ft} - 268{,}549.70\ \text{ft})^2\end{array}} \\ &= 78.09\ \text{ft} \quad (78.1\ \text{ft}) \end{aligned}$$

The answer is (C).

29. Option C is true. Space mean speed is always less than or equal to time mean speed.

Option A is false. The average daily traffic (ADT) is the average of 24 hr counts collected over a number of days greater than one but less than 365. The given definition is for the annual average daily traffic (AADT).

Option B is false because traffic control devices such as traffic lights and stop signs usually cause fixed delay on roadways. The causes of operational delay include traffic side frictions.

Option D is false. The first part of the statement is correct (local roads have more access than mobility), but the second part of the statement is false because local roads nationwide carry a much smaller fraction (only about 20%) of the travel volume.

The answer is (C).

30. The stopping sight distance, S, is the sum of brake reaction and brake distance. From the *AASHTO Green Book*, the brake reaction for 45 mph speed is 165.4 ft with 2.5 sec perception reaction time.

The brake distance for a highway on grade, G, is also given by the *AASHTO Green Book*. The deceleration, a, is assumed to be 11.2 ft/sec² as a default value.

$$\begin{aligned} s_b &= \frac{v^2}{30(f \pm G)} = \frac{v^2}{30\left(\dfrac{a}{g} \pm G\right)} \\ &= \frac{\left(45\ \dfrac{\text{mi}}{\text{hr}}\right)^2}{(30)\left(\dfrac{11.2\ \dfrac{\text{ft}}{\text{sec}^2}}{32.2\ \dfrac{\text{ft}}{\text{sec}^2}} - 0.05\right)} \\ &= 226.6\ \text{ft} \\ S &= 165.4\ \text{ft} + 226.6\ \text{ft} \\ &= 392\ \text{ft} \quad (390\ \text{ft}) \end{aligned}$$

The answer can also be obtained by interpolation from exhibits from the *AASHTO Green Book*.

The answer is (C).

31. Based on the given traffic volumes, link AB represents a major highway, and link CD is a minor road. The traffic is predominantly from A to B and B to A. The turning movements can be accommodated by at-grade turns. A diamond interchange can handle the large volumes between A and B and can incorporate the at-grade turns. A simple diamond interchange is the most suitable.

The answer is (C).

32. Use the average end area method to calculate the volume of cut and the volume of fill.

$$\begin{aligned} V &= \frac{L(A_1 + A_2)}{2} \\ V_{cut} &= \frac{(100\ \text{ft})(9\ \text{ft}^2 + 15\ \text{ft}^2)}{(2)\left(27\ \dfrac{\text{ft}^3}{\text{yd}^3}\right)} \\ &= 44.44\ \text{yd}^3 \\ V_{fill} &= \frac{(100\ \text{ft})(24\ \text{ft}^2 + 32\ \text{ft}^2)}{(2)\left(27\ \dfrac{\text{ft}^3}{\text{yd}^3}\right)} \\ &= 103.70\ \text{yd}^3 \end{aligned}$$

The net earth volume (NEV) is

$$NEV = V_{cut} - V_{fill}$$
$$= 44.44 \text{ yd}^3 - 103.70 \text{ yd}^3$$
$$= -59.26 \text{ yd}^3 \quad (60 \text{ yd}^3 \text{ of fill})$$

The answer is (B).

33. The mixed liquor volatile suspended solids (MLVSS) is

$$\overline{X} = PC$$

C is the concentration of mixed liquor suspended solids (MLSS). P is the percent volatile solids.

$$\overline{X} = PC = \left(0.75 \frac{\text{MLVSS}}{\text{MLSS}}\right)\left(2500 \frac{\text{mg}}{\text{L}} \text{ MLSS}\right)$$
$$= 1875 \text{ mg/L MLVSS}$$

The influent biodegradable COD is

$$S_i = \text{influent COD} - \text{nonbiodegradable COD}$$
$$= 1800 \frac{\text{mg}}{\text{L}} - 110 \frac{\text{mg}}{\text{L}}$$
$$= 1690 \text{ mg/L}$$

The effluent biodegradable COD is

$$S_t = \text{effluent COD} - \text{nonbiodegradable COD}$$
$$= 200 \frac{\text{mg}}{\text{L}} - 110 \frac{\text{mg}}{\text{L}}$$
$$= 90 \text{ mg/L}$$

The hydraulic retention time is

$$\theta = \frac{S_i - S_t}{\overline{X}U}$$

The specific substrate utilization is related to the substrate utilization rate constant by

$$U = k_A S_t$$

Therefore, the hydraulic retention time is

$$\theta = \frac{S_i - S_t}{\overline{X}k_A S_t} = \frac{\left(1690 \frac{\text{mg}}{\text{L}} - 90 \frac{\text{mg}}{\text{L}}\right)\left(1000 \frac{\text{mg}}{\text{g}}\right)}{\left(1875 \frac{\text{mg}}{\text{L}}\right)\left(0.6 \frac{\text{L}}{\text{g·h}}\right)\left(90 \frac{\text{mg}}{\text{L}}\right)}$$
$$= 15.8 \text{ h}$$

(The hydraulic retention time, θ, and the mean cell resistance time, θ_c, are not the same.)

The reactor volume is

$$V = Q\theta$$
$$= \frac{\left(8 \times 10^6 \frac{\text{gal}}{\text{day}}\right)(15.8 \text{ hr})}{\left(7.48 \frac{\text{gal}}{\text{ft}^3}\right)\left(24 \frac{\text{hr}}{\text{day}}\right)}$$
$$= 704,100 \text{ ft}^3 \quad (700,000 \text{ ft}^3)$$

The answer is (D).

34. The unit volume produced from 1.0 in runoff from the watershed is

$$V_{\text{unit}} = A_d h$$
$$= \frac{(43 \text{ mi}^2)\left(640 \frac{\text{ac}}{\text{mi}^2}\right)(1 \text{ in})}{12 \frac{\text{in}}{\text{ft}}}$$
$$= 2293 \text{ ac-ft} \quad (\text{per inch runoff})$$

Since the actual flood hydrograph volume is 3260 ac-ft, the depth of runoff in the flood hydrograph is

$$d = \frac{V_{\text{actual}}}{V_{\text{unit}}}$$
$$= \frac{3260 \text{ ac-ft}}{2293 \frac{\text{ac-ft}}{\text{in}}}$$
$$= 1.42 \text{ in}$$

The actual flood hydrograph peak is 9300 ft³/sec, so the unit hydrograph peak is

$$Q_{p,\text{unit}} = \frac{Q_{p,\text{actual}}}{d}$$
$$= \frac{9300 \frac{\text{ft}^3}{\text{sec}}}{1.42 \text{ in}}$$
$$= 6549 \text{ ft}^3/\text{in-sec} \quad (6500 \text{ ft}^3/\text{in-sec})$$

The answer is (D).

35. The coefficient of velocity is the ratio of actual velocity to the theoretical velocity at the nozzle.

$$C_v = \frac{\text{actual velocity}}{\text{theoretical velocity}} = \frac{v_o}{v_t}$$

The theoretical velocity is

$$v_t = v_2 = \frac{v_o}{C_v}$$

$$= \frac{98 \ \dfrac{\text{ft}}{\text{sec}}}{0.95}$$

$$= 103.2 \text{ ft/sec}$$

The ratio of the *vena contracta* (jet) area to the area of the nozzle is the coefficient of contraction.

$$C_c = \frac{A_{\text{jet}}}{A_2} = 0.8$$

$$A_{\text{jet}} = 0.8 A_2$$

From the continuity equation,

$$Q_1 = Q_2$$

$$A_1 v_1 = A_{\text{jet}} v_2$$

$$v_1 = \frac{0.8 A_2 v_2}{A_1} = \frac{0.8 D_2^2 v_2}{D_1^2}$$

$$= \frac{(0.8)(0.8 \text{ in})^2 \left(103.2 \ \dfrac{\text{ft}}{\text{sec}}\right)}{(3.15 \text{ in})^2}$$

$$= 5.33 \text{ ft/sec}$$

The Bernoulli equation from point 1 (base) to point 2 (jet) is

$$\frac{p_1 g_c}{\rho g} + \frac{v_1^2}{2g} + \cancel{z_1} = \cancel{\frac{p_2 g_c}{\rho g}} + \frac{v_2^2}{2g} + \cancel{z_2}$$

At 160°F, the density of water, ρ, is 61.0 lbm/ft³. The pressure at point 1 is found from

$$\frac{p_1 \left(32.2 \ \dfrac{\text{lbm-ft}}{\text{lbf-sec}^2}\right)}{\left(61.0 \ \dfrac{\text{lbm}}{\text{ft}^3}\right) g} + \frac{\left(5.33 \ \dfrac{\text{ft}}{\text{sec}}\right)^2}{2g} + 0$$

$$= 0 + \frac{\left(103.2 \ \dfrac{\text{ft}}{\text{sec}}\right)^2}{2g} + 0$$

$$p_1 = 10{,}061 \text{ lbf/ft}^2$$

$$p_1 = \frac{10{,}061 \ \dfrac{\text{lbf}}{\text{ft}^2}}{\left(12 \ \dfrac{\text{in}}{\text{ft}}\right)^2}$$

$$= 69.9 \text{ lbf/in}^2 \quad (70 \text{ psi})$$

The answer is (C).

36. The Manning coefficient, n, for brick is 0.016. The cross-sectional area for a rectangular channel is

$$A = dw = (5 \text{ ft})(10 \text{ ft})$$

$$= 50 \text{ ft}^2$$

The wetted perimeter is

$$P = 2d + w = (2)(5 \text{ ft}) + 10 \text{ ft}$$

$$= 20 \text{ ft}$$

The hydraulic radius is

$$R = \frac{A}{P} = \frac{50 \text{ ft}^2}{20 \text{ ft}}$$

$$= 2.5 \text{ ft}$$

The slope is

$$S = \left(\frac{Qn}{1.49 A R^{2/3}}\right)^2$$

$$= \left(\frac{\left(1060 \ \dfrac{\text{ft}^3}{\text{sec}}\right)(0.016)}{(1.49)(50 \text{ ft}^2)(2.5 \text{ ft})^{2/3}}\right)^2$$

$$= 0.01527 \quad (1.5\%)$$

The answer is (B).

37. The NRCS curve numbers are obtained from tables describing standard ground cover types.

unit	curve number
1	39
2	72
3	98

The weighted curve number for the basin is

$$CN_{\text{wt}} = \frac{CN_1 A_1 + CN_2 A_2 + CN_3 A_3}{A_1 + A_2 + A_3}$$

$$= \frac{\begin{array}{c}(39)(200{,}000 \text{ ft}^2) + (72)(300{,}000 \text{ ft}^2) \\ + (98)(180{,}000 \text{ ft}^2)\end{array}}{200{,}000 \text{ ft}^2 + 300{,}000 \text{ ft}^2 + 180{,}000 \text{ ft}^2}$$

$$= 69$$

The storage capacity is

$$S = \frac{1000}{CN_{\text{wt}}} - 10 = \frac{1000}{69} - 10$$

$$= 4.49 \text{ in} \quad (4.5 \text{ in})$$

The answer is (D).

38. For plain sedimentation, type I settling occurs. Assume Stokes' law applies. The settling velocity is

$$v_s = \frac{(SG_{particle} - 1)D_{ft}^2 g}{18\nu}$$

$$= \frac{(2.2 - 1)(6.5 \times 10^{-5} \text{ ft})^2 \left(32.2 \frac{\text{ft}}{\text{sec}^2}\right)}{(18)\left(0.826 \times 10^{-5} \frac{\text{ft}^2}{\text{sec}}\right)}$$

$$= 0.0011 \text{ ft/sec}$$

The Reynolds number is

$$Re = \frac{v_s D}{\nu} = \frac{\left(0.0011 \frac{\text{ft}}{\text{sec}}\right)(6.5 \times 10^{-5} \text{ ft})}{0.826 \times 10^{-5} \frac{\text{ft}^2}{\text{sec}}}$$

$$= 0.0086$$

Since the Reynolds number is less than 1, Stokes' law applies. For 100% removal, the settling velocity is the overflow rate (OFR).

$$OFR = 0.0011 \text{ ft}^3/\text{ft}^2\text{-sec}$$

The area of the tank is

$$A = \frac{Q}{OFR} = \frac{18 \frac{\text{ft}^3}{\text{sec}}}{0.0011 \frac{\text{ft}^3}{\text{ft}^2\text{-sec}}}$$

$$= 16,363 \text{ ft}^2 \quad (16,000 \text{ ft}^2)$$

The depth of the tank is

$$h = t_{settling} v_s$$

$$= (2.5 \text{ hr})\left(3600 \frac{\text{sec}}{\text{hr}}\right)\left(0.0011 \frac{\text{ft}}{\text{sec}}\right)$$

$$= 9.9 \text{ ft} \quad (10 \text{ ft})$$

The answer is (D).

39. The average daily flow is based on per capita factors in the absence of site specific data. Common values are 100–125 gal/capita-day.

The average daily domestic flow is

$$Q_{dom} = Q_{cap} P$$

Q_{cap} is the per capita daily domestic flow. P is the number of people (capita).

$$Q_{dom} = \frac{\left(125 \frac{\text{gal}}{\text{capita-day}}\right)(65,000 \text{ capita})}{10^6 \frac{\text{gal}}{\text{MG}}}$$

$$= 8.125 \text{ MGD}$$

The industrial flow is

$$Q_{ind} = Q_{unit} A$$

Q_{unit} is the daily industrial flow per unit area. A is the area contributing to industrial flow.

$$Q_{ind} = \frac{\left(10,000 \frac{\text{gal}}{\text{ac-day}}\right)(200 \text{ ac})}{10^6 \frac{\text{gal}}{\text{MG}}}$$

$$= 2 \text{ MGD}$$

The excess infiltration is

$$Q_{inf} = EDL$$

E is the excess infiltration rate, D is the diameter of pipe, and L is the length of pipe.

$$Q_{inf} = \left(\frac{600 \frac{\text{gal}}{\text{day-in-mi}}}{10^6 \frac{\text{gal}}{\text{MG}}}\right)\left(\begin{array}{l}(24 \text{ in})(5 \text{ mi}) \\ + (12 \text{ in})(16 \text{ mi}) \\ + (8 \text{ in})(22 \text{ mi})\end{array}\right)$$

$$= 0.293 \text{ MGD}$$

The total annual average daily flow is

$$Q_{total} = Q_{dom} + Q_{ind} + Q_{inf}$$

$$= 8.125 \text{ MGD} + 2.0 \text{ MGD} + 0.293 \text{ MGD}$$

$$= 10.418 \text{ MGD}$$

The design peak daily flow is

$$Q_{peak} = Q_{total}\left(\frac{200\%}{100\%}\right)$$

$$= (10.418 \text{ MGD})(2.0)$$

$$= 20.8 \text{ MGD} \quad (20 \text{ MGD})$$

The answer is (C).

40. The reactions are

$$Cl_2 + H_2O \rightleftharpoons HOCl + H^+ + Cl^-$$

$$SO_2 + HOCl + H_2O \rightleftharpoons Cl^- + SO_4^{-2} + 3H^+$$

One mole of Cl_2 produces one mole of $HOCl$, which reacts with one mole of SO_2. The dose of SO_2 is

$$D_{SO_2} = \left(6 \ \frac{mg}{L} \ Cl_2\right)\left(\frac{1 \ mol \ Cl_2}{71 \ g \ Cl_2}\right)\left(1 \ \frac{mol \ HOCl}{mol \ Cl_2}\right)$$
$$\times \left(1 \ \frac{mol \ SO_2}{mol \ HOCl}\right)\left(64 \ \frac{g \ SO_4}{mol \ SO_4}\right)$$
$$= 5.41 \ mg/L$$

The mass per day of SO_2 is

$$\dot{m} = CQ = D_{SO_2} Q$$
$$= \left(5.41 \ \frac{mg}{L}\right)\left(80 \ \frac{L}{s}\right)\left(86\,400 \ \frac{s}{d}\right)\left(\frac{1 \ kg}{10^6 \ mg}\right)$$
$$= 37.4 \ kg/d \quad (40 \ kg/d)$$

The answer is (C).

Solutions
Afternoon Session Construction

41. The cross section at sta 13+00 is known as a *one-level section* because the surface is at one elevation at all points. The cross section at sta 15+00 is known as a *five-level section* because the surface has five different elevation points.

Excavation volumes can be calculated using either the average end area method or the prismoidal method. When the cross-section shapes are irregular and largely spaced as they are in this problem, contractors often use the average end area method because it results in a larger calculated volume. (Contractors are generally paid by volume.)

The end area at sta 13+00 is

$$A_{\text{sta }13+00} = h(sh + b)$$
$$= (12 \text{ ft})\big((2)(12 \text{ ft}) + 46 \text{ ft}\big)$$
$$= 840 \text{ ft}^2$$

The end area at sta 15+00 is

$$A_{\text{sta }15+00} = \frac{hb + gl + ir}{2}$$
$$= \frac{\begin{array}{c}(9 \text{ ft})(44 \text{ ft}) + (13 \text{ ft})(25 \text{ ft}) \\ + (11 \text{ ft})(28 \text{ ft})\end{array}}{2}$$
$$= 514.5 \text{ ft}^2$$
$$L = 15+00 \text{ sta} - 13+00 \text{ sta}$$
$$= 200 \text{ ft}$$
$$V = \frac{L(A_{\text{sta }13+00} + A_{\text{sta }15+00})}{2}$$
$$= \frac{(200 \text{ ft})(840 \text{ ft}^2 + 514.5 \text{ ft}^2)}{(2)\left(27 \dfrac{\text{ft}^3}{\text{yd}^3}\right)}$$
$$= 5017 \text{ yd}^3 \quad (5000 \text{ yd}^3)$$

The answer is (D).

42. *Planimeters* are mechanical or digital instruments that are used to measure areas enclosed by boundaries such as curved lines, contour lines, and so on.

The volume of material enclosed between the contour lines is

$$V = (\text{contour elevation difference})(\text{average end area})$$
$$= (707 \text{ ft} - 700 \text{ ft})\left(\frac{1800 \text{ ft}^2 + 1200 \text{ ft}^2}{2}\right)$$
$$= 10{,}500 \text{ ft}^3 \quad (11{,}000 \text{ ft}^3)$$

The answer is (B).

43. In differential leveling, only one foresight measurement is taken, but in profile leveling, as many intermediate foresight (IFS) measurements as needed are taken. In the elevation calculations, two equations are used.

$$\text{HI} = \text{elev} + \text{BS}$$
$$\text{elev} = \text{HI} - \text{FS}$$

BS is the backsight, HI is the height of instrument, and FS is the foresight. The instrument is not moved between benchmarks and turning points. Calculate the elevation of each station in series.

$$\text{HI}_{\text{BM }145} = \text{elev}_{\text{BM }145} + \text{BS}_{\text{BM }145}$$
$$= 617.24 \text{ ft} + 3.57 \text{ ft}$$
$$= 620.81 \text{ ft}$$
$$\text{elev}_{\text{sta }1+00} = \text{HI} - \text{IFS}_{\text{sta }1+00}$$
$$= 620.81 \text{ ft} - 4.29 \text{ ft}$$
$$= 616.52 \text{ ft}$$
$$\text{elev}_{\text{sta }2+00} = \text{HI} - \text{IFS}_{\text{sta }2+00}$$
$$= 620.81 \text{ ft} - 5.17 \text{ ft}$$
$$= 615.64 \text{ ft}$$

$$\text{elev}_{\text{TP1}} = \text{HI} - \text{FS}_{\text{TP1}}$$
$$= 620.81 \text{ ft} - 8.21 \text{ ft}$$
$$= 612.60 \text{ ft}$$

$$\text{HI}_{\text{TP1}} = \text{elev}_{\text{TP1}} + \text{BS}_{\text{TP1}}$$
$$= 612.60 \text{ ft} + 2.65 \text{ ft}$$
$$= 615.25 \text{ ft}$$

$$\text{elev}_{\text{sta 3+90}} = \text{HI} - \text{IFS}_{\text{sta 3+90}}$$
$$= 615.25 \text{ ft} - 3.75 \text{ ft}$$
$$= 611.50 \text{ ft} \quad (612 \text{ ft})$$

station	BS	HI	FS	IFS	elevation
BM 145	3.57	**620.81**			617.24
1+00		**620.81**		4.29	**616.52**
2+00		**620.81**		5.17	**615.64**
TP1	2.65	**615.25**	8.21		**612.60**
3+90		**615.25**		3.75	**611.50**

The answer is (C).

44. The bid volume, $V_{\text{footings,total bid}}$, is calculated using the following series of calculations.

$$V_{\text{footing}} = W_{\text{footing}} D_{\text{footing}} L_{\text{footing}}$$

$$V_{\text{footings,continuous}} = \frac{(16 \text{ in})(8 \text{ in})(207 \text{ ft})}{\left(12 \frac{\text{in}}{\text{ft}}\right)^2 \left(27 \frac{\text{ft}^3}{\text{yd}^3}\right)}$$
$$= 6.8 \text{ yd}^3$$

$$V_{\text{footings,column}} = (\text{number of footings}) A_{\text{footings}} D_{\text{footing}}$$
$$= \frac{(4)(3 \text{ ft})(3 \text{ ft})(1 \text{ ft})}{27 \frac{\text{ft}^3}{\text{yd}^3}}$$
$$= 1.3 \text{ yd}^3$$

$$V_{\text{footings,total bid}} = 6.8 \text{ yd}^3 + 1.3 \text{ yd}^3$$
$$= 8.1 \text{ yd}^3$$

The actual constructed volume is found similarly.

$$V_{\text{footings,continuous}} = \frac{(16 \text{ in})(8 \text{ in})(207 \text{ ft})(1.5)}{\left(12 \frac{\text{in}}{\text{ft}}\right)^2 \left(27 \frac{\text{ft}^3}{\text{yd}^3}\right)}$$
$$= 10.2 \text{ yd}^3$$

$$V_{\text{footings,column}} = \frac{(4)(3 \text{ ft})(3 \text{ ft})(1 \text{ ft})(1.5)}{27 \frac{\text{ft}^3}{\text{yd}^3}}$$
$$= 2.00 \text{ yd}^3$$

$$V_{\text{footings,total constructed}} = 10.2 \text{ yd}^3 + 2.00 \text{ yd}^3$$
$$= 12.2 \text{ yd}^3$$

$$\text{additional concrete needed} = V_{\text{footings,total constructed}}$$
$$- V_{\text{footings,total bid}}$$
$$= 12.2 \text{ yd}^3 - 8.1 \text{ yd}^3$$
$$= 4.1 \text{ yd}^3$$

The answer is (B).

45. The mass diagram rises from sta 0+00 to sta 6+00, which means that a portion of the ground profile must be an excavation. Then, the mass diagram curve falls from sta 6+00 to sta 12+00, which means that a portion of the ground profile is an embankment. Peaks occur at a transition from excavation to embankment. The diagram that has these characteristics is option B.

The answer is (B).

46. The length of Skyline Road is

$$32+00 \text{ sta} - 22+00 \text{ sta} = 10 \text{ sta} \quad (1000 \text{ ft})$$

110 lbf/yd^2 is the given area for a 1 in thick asphalt section of AC pavement, and the asphalt is 6 in thick.

$$\text{AC tonnage} = \frac{(6 \text{ in})\left(110 \frac{\text{lbf}}{\text{yd}^2\text{-in}}\right)(1000 \text{ ft})(40 \text{ ft})}{\left(3 \frac{\text{ft}}{\text{yd}}\right)^2 \left(2000 \frac{\text{lbf}}{\text{ton}}\right)}$$
$$= 1467 \text{ tons} \quad (1470 \text{ tons})$$

3000 lbf/yd^3 is the given volume for the AB pavement, and the asphalt is 6 in thick.

$$\text{AB tonnage} = \frac{(1000 \text{ ft})(40 \text{ ft})(6 \text{ in})\left(3000 \frac{\text{lbf}}{\text{yd}^3}\right)}{\left(12 \frac{\text{in}}{\text{ft}}\right)\left(27 \frac{\text{ft}^3}{\text{yd}^3}\right)\left(2000 \frac{\text{lbf}}{\text{ton}}\right)}$$
$$= 1111 \text{ tons} \quad (1110 \text{ tons})$$

The answer is (A).

47. The total cost, C_{total}, is the sum of the cost of time, C_{time}, and the cost of materials, $C_{materials}$.

$$C_{time} = \frac{AC_{labor} \times (\text{labor overhead multiplier})}{\text{production rate}}$$

$$= \frac{(34{,}600 \text{ ft}^2)\left(25 \dfrac{\$}{\text{hr}}\right)(1.35)}{700 \dfrac{\text{ft}^2}{\text{hr}}}$$

$$= \$1668$$

$$C_{materials} = \frac{AC_{paint\,bucket} \times (\text{materials overhead multiplier})}{\text{total paint coverage per bucket}}$$

$$= \frac{(34{,}600 \text{ ft}^2)(\$10)(1.30)}{1200 \text{ ft}^2}$$

$$= \$375$$

$$C_{total} = \$1668 + \$375$$

$$= \$2043 \quad (\$2040)$$

The answer is (C).

48. The cost, C, of all bays is calculated by multiplying the number of bays by the cost of one bay. The total cost of construction is the sum of all costs associated with the construction.

$$C_{type\,A\,bays} = (6)(\$100{,}000)$$

$$= \$600{,}000$$

$$C_{type\,B\,bays} = (6)(\$85{,}000)$$

$$= \$510{,}000$$

$$C_{type\,C\,bays} = (4)(\$70{,}000)$$

$$= \$280{,}000$$

$$C_{all\,bays} = \$600{,}000 + \$510{,}000 + \$280{,}000$$

$$= \$1{,}390{,}000$$

$$C_{total\,construction} = C_{all\,bays} + C_{entrances} + C_{elevator}$$
$$+ \, C_{end\,walls} + C_{equipment}$$

$$= \$1{,}390{,}000 + \$20{,}000 + \$40{,}000$$
$$+ \, \$60{,}000 + \$80{,}000$$

$$= \$1{,}590{,}000$$

The answer is (D).

49. The straight line depreciation method is the most common and easiest method of computing depreciation, D. C is the purchase price, and S_n is the expected salvage value. The IRS has established five years as the useful life for computers.

$$D = \frac{C - S_n}{n}$$

$$= \frac{\$30{,}000 - \$2500}{5 \text{ yr}}$$

$$= \$5500/\text{yr}$$

The firm can deduct \$5500 each year for the next five years from the company's revenue.

The answer is (C).

50. The sum of the years' digits (SOYD) and the double declining balance method are accelerated depreciation methods. The process for calculating the depreciation of the computers using the SOYD method is as follows.

$$D_j = \frac{(C - S_n)(n - j + 1)}{T}$$

D is depreciation; C is cost; S_n is the expected salvage value at year n; j is the number of years since purchase; T is the sum of the digits from 1 to n inclusive.

$$T = 5 + 4 + 3 + 2 + 1$$

$$= 15$$

$$D_1 = \frac{(\$30{,}000 - \$2500)(5 - 1 + 1)}{15}$$

$$= \$9167$$

$$D_2 = \frac{(\$30{,}000 - \$2500)(5 - 2 + 1)}{15}$$

$$= \$7333$$

$$D_{1+2} = \$9167 + \$7333$$

$$= \$16{,}500 \quad (\$17{,}000)$$

Note that the straight line (SL) method only gives a fixed \$5500 per year, or \$11,000 for two years. Therefore, SOYD gives higher depreciation value in the early years of owning computers or equipment. The owner of this firm, by choosing SOYD method, is able to deduct \$16,500 for the depreciation of the computers, while an owner using the SL method would deduct only \$11,000 for the same two-year period.

The answer is (C).

51. EBITDA is an acronym for earnings before interest, taxes, depreciation, and amortization, and is a gross measure of a company's financial performance. It is not the company's cash flow, net income, or profit. For this reason, if two companies have the same EBITDA, they may not be equally profitable. Interest, tax, depreciation, and so on, have an impact on profitability. EBITDA alone is not a good measure of financial comparison. Other information is needed, such as debt ratios, backlog of orders, and labor cost.

The answer is (A).

52. According to OSHA 29 CFR 1926, Subpart P, App. B, Fig. B-1, (1) "simple slope — short term," from *OSHA Standards for the Construction Industry*, $1/2$:1 (H:V) is the maximum allowable slope for a temporary excavation that is 12 ft high or less, has a simple slope, and is open 24 hr or less.

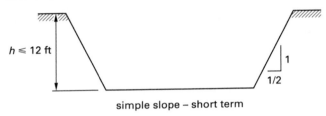

simple slope – short term

The answer is (A).

53. An experience modification factor is a ratio that insurers use to compare the losses of those insured to the losses of others in a similar industry and location for the previous three years. In particular, this is an adjustment that applies to workers' compensation insurance.

The calculations are made by the National Council on Compensation Insurance (NCCI), so if a firm has more workers' claims in one year than in previous years, it will only count against the firm if the claims exceed average industry claims. In essence, safety management of similar firms in the same locale is being compared in a given three-year policy period.

The answer is (D).

54. Use the U.S. Department of Labor, Bureau of Labor Statistics formula for calculating incidence rate, *I*. The formula assumes 100 employees working 40 hours per week for 50 weeks, or 200,000 labor hours.

$$I = \frac{(\text{number of injuries and illnesses})(200{,}000 \text{ hr})}{\text{employee hours worked}}$$
$$= \frac{(5)(200{,}000 \text{ hr})}{80{,}000 \text{ hr}}$$
$$= 12.5 \quad (13)$$

The answer is (C).

55. Pile load tests are done to verify computed design capacities. The pile load tests are run until failure occurs in the pile or twice the design load is achieved with a $1/4$ in settlement in at least 60 hr.

The PL/AE line in the diagram represents the shortening of the pile magnitude. The main concern is the failure criterion line, which intersects the load-displacement curve at approximately 280 kips. This is the ultimate load. The factor of safety, FS, is given as 2.0, so the allowable load is

$$Q_a = \frac{P}{\text{FS}}$$
$$= \frac{280 \text{ kips}}{2.0}$$
$$= 140 \text{ kips}$$

The answer is (B).

56. Note that in the problem's illustration the forces are given per foot of wall length.

$$\text{FS}_{\text{OT}} = \frac{M_{\text{resisting}}}{M_{\text{overturning}}} = \frac{W x}{P_a b}$$
$$= \frac{\left(3200 \frac{\text{lbf}}{\text{ft}}\right)(3 \text{ ft})}{\left(1700 \frac{\text{lbf}}{\text{ft}}\right)(3 \text{ ft})}$$
$$= 1.88 \quad (1.9)$$

Normally, factor of safety against overturning should exceed 1.5. Due to the large lateral movement required to mobilize passive resistance, rotational failure may occur prior to sufficient wall translation.

Passive resistance may not prevent overturning if soil cover at the toe is minimal or if soil in front of the wall has been excavated, slopes downward, has loosened, or could develop tension cracks.

The answer is (C).

57. It is more conservative to ignore adhesion at the footing base due to possible saturation of soils and consequent loss of cohesion. Without consideration for adhesion at the base of the footings, the factor of safety against sliding is

$$\text{FS}_{\text{SL}} = \frac{P_p + W \tan \delta}{P_a}$$

δ is $2/3$ the soil friction angle, ϕ.

$$\delta = \tfrac{2}{3}\phi = \left(\tfrac{2}{3}\right)(33°) = 22°$$

$$\text{FS}_{\text{SL}} = \frac{400 \frac{\text{lbf}}{\text{ft}} + \left(3200 \frac{\text{lbf}}{\text{ft}}\right)(\tan 22°)}{1700 \frac{\text{lbf}}{\text{ft}}} = 0.996 \quad (1.0)$$

FS_{SL} should be greater than 1.5. Increasing footing embedment, providing a keyway, or replacing foundation soils with aggregate base would all increase the FS against sliding.

The answer is (A).

58. In order to ensure that each member is strong enough to transfer the loads to the next member, the load path for the given illustration must come from the stud to the strap to the beam to the column to the footing and, finally, to the soil.

The answer is (D).

59. The Swedish chemist Albert Atterberg discovered that cohesive materials are best classified on their plastic and liquid limits. (See ASTM D4318.) The plasticity index (PI) is calculated by subtracting the plastic limit (PL) from the liquid limit (LL).

Rolling action is used to determine the PL of the soils. Only samples with excess moisture can be rolled to $1/16$ in without breaking, so PL is artificially increased above the true PL (at $1/8$ in). Evaporation during the extended rolling time will decrease the moisture content slightly, but moisture content must still exceed the true PL (at $1/8$ in) in order to roll to $1/16$ in. Therefore, PL is increased, thereby decreasing the PI, which becomes apparent once PL is subtracted from LL.

The answer is (A).

60. Soluble sulfates content is the controlling factor in this problem. According to Table 2-2 of *Design and Control of Concrete Mixtures*, type V cement should be specified in this instance. This table has been adopted by various building codes, including the IBC.

In addition, in order to achieve low permeability, a water/cement ratio of less than 0.45 is required. A low water/cement ratio can also help achieve a 4500 psi compressive strength.

The answer is (D).

61. 29 CFR Part 1926, Subpart P classifies soil into four categories: types A, B, C, and layered geological strata. Type C-60 is commonly used to describe a soil that shows signs of cohesiveness, is not flowing or submerged, but does not fit into type A or type B classifications. Type C-60 is a classification of the Trench Shoring and Shielding Association (TSSA) and is not an OSHA classification.

The answer is (C).

62. The four most common methods of welding are shielded metal arc welding (SMAW), gas metal arc welding (GMAW), submerged arc welding (SAW), and flux core arc welding (FCAW).

The answer is (B).

63. Compaction production for an 825H sheepsfoot roller is given in the following formula, with W as width of one wheel (multiplying by 2 gives the compacted width), L as the compacted lift thickness in inches, and S as the speed in miles per hour.

$$\begin{aligned} \text{compaction} \atop \text{production} &= \frac{(16.3)(2)\,WLS}{\text{no. of machine passes}} \\ &= \frac{(16.3)(2)(3.7\text{ ft})(9\text{ in})(7\text{ mph})}{5\text{ passes}} \\ &= 1520\text{ yd}^3/\text{hr} \quad (1500\text{ yd}^3/\text{hr}) \end{aligned}$$

The answer is (D).

64. Actual production, P_{actual}, is calculated from F_p, the number of passes, F_s, the speed, and F_t, the thickness.

$$\begin{aligned} P_{actual} &= \frac{P_{assumed}\left(\dfrac{F_{s,actual}}{F_{s,assumed}}\right)\left(\dfrac{F_{p,assumed}}{F_{p,actual}}\right)\left(\dfrac{F_{t,actual}}{F_{t,assumed}}\right)}{1\text{ hr}} \\ &= \frac{(480\text{ yd}^3)\left(\dfrac{5\text{ mph}}{6\text{ mph}}\right)\left(\dfrac{7}{6}\right)\left(\dfrac{4\text{ in}}{5\text{ in}}\right)}{1\text{ hr}} \\ &= 373\text{ yd}^3/\text{hr} \quad (370\text{ yd}^3/\text{hr}) \end{aligned}$$

The production rate decreases with increases in the number of passes.

The answer is (B).

65. The required grade horsepower is

$$\begin{aligned} \text{required} \atop \text{grade} \atop \text{horsepower} &= \text{GMW}(\text{grade} + \text{roll resistance})(\text{speed}) \\ &= \frac{(120\text{ tons})\left(2000\ \dfrac{\text{lbf}}{\text{ton}}\right)}{\left(550\ \dfrac{\text{ft-lbf}}{\text{hp-sec}}\right)\left(3600\ \dfrac{\text{sec}}{\text{hr}}\right)\left(5280\ \dfrac{\text{ft}}{\text{mi}}\right)} \\ &\quad \times (0.07+0.03)(10\text{ mph}) \\ &= 640\text{ hp} \end{aligned}$$

The answer is (A).

66. Rim pull (which is also a company name) is the tractive force at the tire-ground contact point.

$$\text{rim pull} = \frac{(\text{efficiency})(\text{hp})}{\text{speed}}$$

$$= \frac{(0.80)(300 \text{ hp})\left(\dfrac{\left(550 \dfrac{\text{ft-lbf}}{\text{hp-sec}}\right)\left(3600 \dfrac{\text{sec}}{\text{hr}}\right)}{5280 \dfrac{\text{ft}}{\text{mi}}}\right)}{10 \text{ mph}}$$

$$= 9000 \text{ lbf}$$

The answer is (B).

67. Estimating soil permeability is the most critical aspect in dewatering a site, because adjustments can be made for other aspects on the job site. For example, if the dewatering capacity is slow, more well points can be added, and if settlement of nearby structures is anticipated, settlement references can be installed.

The answer is (D).

68. Assuming that load is unchanged, the lifting capacity of a crane decreases as the load radius and the boom length increase.

The answer is (B).

69. Calculate the costs of the concrete and the rebar needed.

$$\begin{aligned}\text{concrete} \atop \text{cost} &= \left(\frac{(10,000 \text{ ft}^2)(6 \text{ in})}{\left(27 \dfrac{\text{ft}^3}{\text{yd}^3}\right)\left(12 \dfrac{\text{in}}{\text{ft}}\right)}\right)\left(200 \frac{\$}{\text{yd}^3}\right)\\ &= \$37,037 \quad (\$37,000)\end{aligned}$$

The rebar spacing in both directions is

$$s = \frac{9 \text{ in}}{12 \dfrac{\text{in}}{\text{ft}}} = 0.75 \text{ ft}$$

$$\begin{aligned}\text{rebar} \atop \text{cost} &= \left(\begin{array}{l}\left(\dfrac{\text{slab length}}{\text{rebar spacing}} + 1\right)(\text{slab width}) \\ + \left(\dfrac{\text{slab width}}{\text{rebar spacing}} + 1\right)(\text{slab length})\end{array}\right)\\ &\quad \times (\text{unit cost})\\ &= \left(\begin{array}{l}\left(\dfrac{100 \text{ ft}}{0.75 \text{ ft}} + 1\right)(100 \text{ ft}) \\ + \left(\dfrac{100 \text{ ft}}{0.75 \text{ ft}} + 1\right)(100 \text{ ft})\end{array}\right)\left(5 \frac{\$}{\text{linear ft}}\right)\\ &= \$134,333 \quad (\$134,000)\end{aligned}$$

$$\begin{aligned}\text{vapor} \atop \text{retarder} \atop \text{cost} &= \left(6 \frac{\$}{\text{ft}^2}\right)(10,000 \text{ ft}^2) = \$60,000\end{aligned}$$

$$\text{labor cost} = \left(2 \frac{\$}{\text{ft}^2}\right)(10,000 \text{ ft}^2) = \$20,000$$

The total cost of the bid is the sum of the costs of the grand base, form work, concrete, rebar, vapor retarder, and labor, multiplied by 1.07, which is the multiplier for the cost of waste disposal.

$$\begin{aligned}\text{total} \atop \text{estimated} \atop \text{cost} &= (1.07)\left(\begin{array}{l}\$12,000 + \$9000 \\ + \$37,000 + \$134,000 \\ + \$60,000 + \$20,000\end{array}\right)\\ &= \$291,040 \quad (\$291,000)\end{aligned}$$

The answer is (D).

70. The weight per foot of each steel shape is given in the steel designation. Multiply this by the length, the count, and the cost for each steel type and add them together.

$$\begin{aligned}C_{\text{W}10\times22} &= (3)(350 \text{ ft})\left(22 \frac{\text{lbf}}{\text{ft}}\right)\left(7 \frac{\$}{\text{lbf}}\right)\\ &= \$161,700\\ C_{\text{W}18\times65} &= (4)(460 \text{ ft})\left(65 \frac{\text{lbf}}{\text{ft}}\right)\left(10 \frac{\$}{\text{lbf}}\right)\\ &= \$1,196,000\\ C_{\text{W}24\times76} &= (2)(700 \text{ ft})\left(76 \frac{\text{lbf}}{\text{ft}}\right)\left(12 \frac{\$}{\text{lbf}}\right)\\ &= \$1,276,800\\ C_{\text{total}} &= \$161,700 + \$1,196,000 + \$1,276,800\\ &= \$2,634,500 \quad (\$2,635,000)\end{aligned}$$

The answer is (A).

71. The equivalent uniform annual cost (EUAC) is the average cost of operation, maintenance, and ownership over all the years up to a given point. In general, equipment maintenance is low in the early years and salvage value is high. As years pass, maintenance costs increase and salvage values decline. Equipment should be replaced when the EUAC value begins to rise.

In this case, the EUAC rises if the equipment is kept for three years, so the equipment should be replaced at the end of the second year.

The answer is (A).

72. The building's periphery length is

$$L = 50 \text{ ft} + 50 \text{ ft} + 80 \text{ ft} + 80 \text{ ft} = 260 \text{ ft}$$

$$V_{\text{excavation}} = D_{\text{excavation}} A_{\text{building}}$$

$$+ L_{\text{building}} D_{\text{excavation}} w_{\text{trench}}$$

$$\times (\text{H:V slope}_{\text{excavation}})$$

$$= \frac{\begin{array}{c}(10 \text{ ft})(50 \text{ ft})(80 \text{ ft}) \\[4pt] + (260 \text{ ft})\left(\dfrac{1}{2}\right)(10 \text{ ft})(5 \text{ ft})\end{array}}{27 \dfrac{\text{ft}^3}{\text{yd}^3}}$$

$$= 1722 \text{ yd}^3 \quad (1700 \text{ yd}^3)$$

(1740 yd^3 if the average perimeter length or length of 280 ft is used. 1734 yd^3 if the frustrum equation is used.)

The answer is (B).

73. The inflation rate affects the purchasing power, but it does not affect the accumulation rate. The original principal is

$$P_{\text{original}} = \frac{P_{\text{current}}}{(1+i)^n} = \frac{\$50{,}000}{(1+0.08)^3}$$

$$= \$39{,}692 \quad (\$39{,}700)$$

The answer is (D).

74. To double an investment, find the F/P value nearest to 2 in an engineering economics table for $I = 12\%$.

The value is 1.9738 (nearly double) at 6 yr.

To triple an investment, find the F/P value nearest to 3.

The value is 3.1058 (nearly triple) at 10 yr.

The answer is (D).

75. The duration, D, of an activity is calculated by dividing the quantity of work by the productivity. The total duration, D_{total}, is the sum of time spent on all activities.

$$D_{\text{earthwork}} = \frac{600 \text{ hr}}{12 \dfrac{\text{hr}}{\text{day}}} = 50 \text{ days}$$

$$D_{\text{surveying}} = \frac{60 \text{ hr}}{10 \dfrac{\text{hr}}{\text{day}}} = 6 \text{ days}$$

$$D_{\text{design phase}} = \frac{500 \text{ hr}}{10 \dfrac{\text{hr}}{\text{day}}} = 50 \text{ days}$$

$$D_{\text{permitting}} = \frac{100 \text{ hr}}{5 \dfrac{\text{hr}}{\text{day}}} = 20 \text{ days}$$

$$D_{\text{total}} = D_{\text{earthwork}} + D_{\text{surveying}}$$

$$+ D_{\text{design phase}} + D_{\text{permitting}}$$

$$= 50 \text{ days} + 6 \text{ days} + 50 \text{ days} + 20 \text{ days}$$

$$= 126 \text{ days} \quad (130 \text{ days})$$

The answer is (B).

76. CV is the cost variance, calculated by subtracting ACWP from BCWP. PCV is the percent cost variance, calculated by dividing CV by BCWP.

$$\text{PCV} = \frac{\text{BCWP} - \text{ACWP}}{\text{BCWP}}$$

$$= \left(\frac{\$400{,}000 - \$380{,}000}{\$400{,}000}\right) \times 100\%$$

$$= 5\%$$

Since this PCV is a positive variance, the project is under budget.

In the beginning of a construction project, project managers are concerned primarily with scheduling, cost, and quality. After those items are under control, firms desire to control time and cost together. This necessitates performance measurements like ACWP, BCWP, and so on.

The answer is (C).

77. Extend a vertical line upward from various points along the horizontal axis to see how many activity lines are crossed. Between days 0 and 10, there are four activities. Between days 10 and 20, there are also four activities. Between days 20 and 30, there are five activities. Between days 40 and 60, there are only two activities.

Therefore, the owners should visit between days 20 and 30.

The answer is (C).

78. The Pareto optimal curve shows the points at which negotiation could be reached. Any point in the space under the curve represents moments during which neither party is interested in reaching an agreement. For this reason, even though both parties are equally interested at point D, neither is interested in agreement, and this point is not optimal in comparison to the others. Both points A and C are on the curve, and during those times, agreement is possible. Point C is the better choice, because the agreement terms will be equally appreciated. While it is unlikely for both parties to be 100% interested in a negotiation, Point B is the best point on the grid, as it represents a point in time during which both parties are equally and maximally interested.

The answer is (B).

79. To lead a 50 person engineering department, the two most important traits are leadership and stress management. Optimism and empathy are good secondary traits. Option D is the only option with the two primary traits scored high.

The answer is (D).

80. When scheduling a project, cost, productivity, and safety must all be considered. Long hours, more equipment, and more people may help construction be completed sooner, but safety may be at risk.

If the owner indicates that the project does not need to be rushed, equipment and people should be minimized. Option A meets the client's preferences using the fewest pieces of equipment.

The answer is (A).

Solutions
Afternoon Session
Geotechnical

81. Consolidation is calculated at point A, the midpoint of the second clay stratum (layer 2). It is reasonable to assume that the increase in pressure at the midpoint of the layer can be computed using the 2:1 stress distribution method, and that the stress spread starts at the lower third point of the piles.

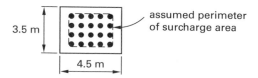

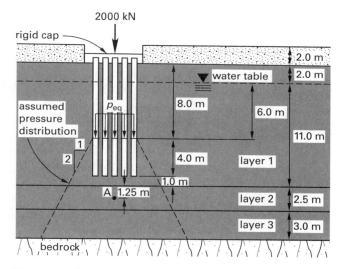

The equivalent pressure at the lower third of the piles, p_{eq}, is

$$p_{eq} = \frac{P}{BL}$$
$$= \frac{2000 \text{ kN}}{(3.5 \text{ m})(4.5 \text{ m})}$$
$$= 127.0 \text{ kN/m}^2$$

To determine the initial pressure, p_o, calculate the pressure from each soil layer individually at the points of interest.

The initial pressure at point A is

$$
\begin{aligned}
p_{o,A} &= \sum \gamma_i z_i \\
&= \left(17.5 \ \frac{\text{kN}}{\text{m}^3}\right)(2 \text{ m}) + \left(19.0 \ \frac{\text{kN}}{\text{m}^3}\right)(2 \text{ m}) \\
&\quad + \left(19.0 \ \frac{\text{kN}}{\text{m}^3} - 9.81 \ \frac{\text{kN}}{\text{m}^3}\right)(11.0 \text{ m}) \\
&\quad + \left(18.0 \ \frac{\text{kN}}{\text{m}^3} - 9.81 \ \frac{\text{kN}}{\text{m}^3}\right)(1.25 \text{ m}) \\
&= 184.3 \text{ kN/m}^2 \quad (184.3 \text{ kPa})
\end{aligned}
$$

To determine the change in pressure, Δp, calculate the pressure change at the midpoint of layer 2 using the 2:1 stress distribution method, and assume that the stress spread starts at the lower third point of the piles. The change in pressure at point A is

$$
\Delta p_A = \left(127.0 \ \frac{\text{kN}}{\text{m}^2}\right) \left(\frac{(3.5 \text{ m})(4.5 \text{ m})}{\begin{array}{c}(4.5 \text{ m} + (2)(6.25 \text{ m})(0.50)) \\ \times \left(\begin{array}{c}3.5 \text{ m} + (2) \\ \times (6.25 \text{ m})(0.50)\end{array}\right)\end{array}} \right)
$$
$$= 19.1 \text{ kN/m}^2 \quad (19.1 \text{ kPa})$$

For a layer of soil with thickness H, the settlement, S, is

$$S = \frac{C_c}{1 + e_o} H \log_{10} \frac{p_o + \Delta p}{p_o}$$

Therefore, the settlement in clay layer 2 is

$$
\begin{aligned}
S_2 &= \frac{C_c}{1 + e_o} H \log_{10} \frac{p_o + \Delta p}{p_o} \\
&= \left(\frac{0.32}{1 + 1.03}\right)(2.5 \text{ m}) \\
&\quad \times \log_{10} \left(\frac{184.3 \ \frac{\text{kN}}{\text{m}^2} + 19.1 \ \frac{\text{kN}}{\text{m}^2}}{184.3 \ \frac{\text{kN}}{\text{m}^2}}\right) \\
&= 0.017 \text{ m} \quad (17 \text{ mm})
\end{aligned}
$$

The answer is (B).

82. The bentonite content is based on dry weight. The dry unit weight is

$$\gamma_d = \frac{\gamma}{1+w} = \frac{17.0 \ \frac{\text{kN}}{\text{m}^3}}{1+0.18}$$
$$= 14.4 \ \text{kN/m}^3$$

Each lift has a thickness, t, of 125 mm. The total layer weight, W_{total}, of each lift is

$$W_{\text{total}} = t\gamma_d = \frac{(125 \ \text{mm})\left(14.4 \ \frac{\text{kN}}{\text{m}^3}\right)}{1000 \ \frac{\text{mm}}{\text{m}}}$$
$$= 1.8 \ \text{kN/m}^2$$

As 8% (by weight) of this mixture is bentonite, the total weight of bentonite required per lift is

$$W_{\text{bentonite}} = 0.08 W_{\text{total}} = (0.08)\left(1.8 \ \frac{\text{kN}}{\text{m}^2}\right)$$
$$= 0.144 \ \text{kN/m}^2 \quad (0.14 \ \text{kN/m}^2)$$

The answer is (A).

83. The hydraulic gradient, i, can be calculated from the leachate head, h, and the thickness of the clay liner, t.

$$i = \frac{H}{t} = \frac{h_1 - h_2}{t} = \frac{0.5 \ \text{m} - 0 \ \text{m}}{1.2 \ \text{m}}$$
$$= 0.417 \ \text{m/m}$$

The effective velocity is given by

$$v_a = Ki = \frac{\left(6 \times 10^{-7} \ \frac{\text{cm}}{\text{s}}\right)\left(0.417 \ \frac{\text{m}}{\text{m}}\right)}{100 \ \frac{\text{cm}}{\text{m}}}$$
$$= 2.50 \times 10^{-9} \ \text{m/s}$$

The flow for the entire landfill on an annual basis is

$$Q = v_a A$$
$$= \left(2.50 \times 10^{-9} \ \frac{\text{m}}{\text{s}}\right)\left(3.16 \times 10^7 \ \frac{\text{s}}{\text{yr}}\right)$$
$$\times (300 \ \text{m})(400 \ \text{m})$$
$$= 9480 \ \text{m}^3/\text{yr} \quad (10^4 \ \text{m}^3/\text{yr})$$

The answer is (B).

84. Find the unit weights of the soils. The dry unit weight of sand is

$$\gamma_d = \frac{\gamma}{1+w} = \frac{18.4 \ \frac{\text{kN}}{\text{m}^3}}{1+0.18}$$
$$= 15.6 \ \text{kN/m}^3$$

The unit weight of sand above the water table (after lowering the water table) is

$$\gamma = \gamma_d(1+w) = \left(15.6 \ \frac{\text{kN}}{\text{m}^3}\right)(1+0.11)$$
$$= 17.3 \ \text{kN/m}^3$$

The dry unit weight of clay is

$$\gamma_d = \frac{\text{SG}\gamma_w}{1+e} = \frac{(2.68)\left(9.81 \ \frac{\text{kN}}{\text{m}^3}\right)}{1+1.15}$$
$$= 12.2 \ \text{kN/m}^3$$

The water content of clay is

$$w = \frac{Se}{\text{SG}} = \frac{(1.0)(1.15)}{2.68}$$
$$= 0.43$$

The saturated unit weight of clay is

$$\gamma_{\text{sat}} = \gamma_d(1+w) = \left(12.2 \ \frac{\text{kN}}{\text{m}^3}\right)(1+0.43)$$
$$= 17.4 \ \text{kN/m}^3$$

The initial effective stress at the midpoint of the clay layer is

$$\bar{\sigma}_o = \sigma - u$$
$$= \left(18.4 \ \frac{\text{kN}}{\text{m}^3} - 9.81 \ \frac{\text{kN}}{\text{m}^3}\right)(30.0 \ \text{m})$$
$$+ \left(17.4 \ \frac{\text{kN}}{\text{m}^3} - 9.81 \ \frac{\text{kN}}{\text{m}^3}\right)(4.0 \ \text{m})$$
$$= 288 \ \text{kN/m}^2 \quad (288 \ \text{kPa})$$

The final effective stress at the midpoint of the clay layer is

$$\bar{\sigma}_f = \sigma - u$$
$$= \left(17.3 \ \frac{\text{kN}}{\text{m}^3}\right)(16.3 \ \text{m}) + \left(18.4 \ \frac{\text{kN}}{\text{m}^3} - 9.81 \ \frac{\text{kN}}{\text{m}^3}\right)$$
$$\times (13.7 \ \text{m}) + \left(17.4 \ \frac{\text{kN}}{\text{m}^3} - 9.81 \ \frac{\text{kN}}{\text{m}^3}\right)(4.0 \ \text{m})$$
$$= 430 \ \text{kN/m}^2 \quad (430 \ \text{kPa})$$

The ultimate settlement of the clay layer is

$$S = \frac{HC_c \log \frac{\overline{\sigma}_f}{\overline{\sigma}_o}}{1+e}$$

$$= \frac{(8.0 \text{ m})(0.32)\left(\log \frac{430 \text{ kPa}}{288 \text{ kPa}}\right)}{1+1.15}$$

$$= 0.21 \text{ m} \quad (0.20 \text{ m})$$

The answer is (B).

85. For a falling-head permeameter, the coefficient of permeability is

$$K = \frac{A'L}{At} \ln \frac{h_i}{h_f}$$

The ratios of the areas can be found from the ratio of the diameters. A is the cross-sectional area of the soil, A' is the cross-sectional area of the standpipe.

$$\frac{A'}{A} = \frac{\pi \left(\frac{d'}{2}\right)^2}{\pi \left(\frac{d}{2}\right)^2} = \left(\frac{d'}{d}\right)^2$$

$$= \left(\frac{0.25 \text{ cm}}{10.0 \text{ cm}}\right)^2$$

$$= 0.000625$$

$$K = \left(\frac{A'}{A}\right)\left(\frac{L}{t}\right) \ln \frac{h_i}{h_f}$$

$$= (0.000625)\left(\frac{6 \text{ cm}}{1278 \text{ s}}\right)\left(\ln \frac{100 \text{ cm}}{50 \text{ cm}}\right)$$

$$= 2 \times 10^{-6} \text{ cm/s}$$

The answer is (D).

86. The time required to achieve 90% consolidation, t_{90}, is

$$t_{90} = \frac{T_{90} H^2}{C_v}$$

This can be rearranged to give

$$T_{90} = \frac{C_v t_{90}}{H^2}$$

For the clayey silt sample,

$$t_{90,s} = (10 \text{ min})\left(60 \frac{\text{s}}{\text{min}}\right) + 46 \text{ s}$$

$$= 646 \text{ s}$$

$$H_s = \frac{\left(\frac{1}{2}\right)(5 \text{ cm})}{100 \frac{\text{cm}}{\text{m}}}$$

$$= 0.025 \text{ m} \quad \text{[double drainage]}$$

$$T_{90} = \frac{C_v t_{90,s}}{H_s^2} = \frac{C_v(646 \text{ s})}{(0.025 \text{ m})^2}$$

For the 25 m clayey silt layer,

$$H_l = \frac{25 \text{ m}}{2} = 12.5 \text{ m} \quad \text{[double drainage]}$$

$$T_{90} = \frac{C_v t_{90,l}}{H_l^2} = \frac{C_v t_{90,l}}{(12.5 \text{ m})^2}$$

Since T_{90} is the same for both the site soil and the sample soil, the right hand sides of the last two equations are also equivalent.

$$\frac{C_v(646 \text{ s})}{(0.025 \text{ m})^2} = \frac{C_v t_{90,l}}{(12.5 \text{ m})^2}$$

$$t_{90,l} = \frac{\frac{(646 \text{ s})(12.5 \text{ m})^2}{(0.025 \text{ m})^2}}{\left(60 \frac{\text{s}}{\text{min}}\right)\left(60 \frac{\text{min}}{\text{h}}\right)\left(24 \frac{\text{h}}{\text{d}}\right)\left(365 \frac{\text{d}}{\text{yr}}\right)}$$

$$= 5.12 \text{ yr} \quad (5 \text{ yr})$$

The answer is (C).

87. For the settlement of 124 mm, the degree of consolidation is

$$U_z = \frac{\Delta H}{\Delta H_{\text{ult}}} = \frac{124 \text{ mm}}{502 \text{ mm}}$$

$$= 0.247$$

The time factor, T_v, can be found in a table of approximate time factors, or calculated from the following equation.

$$T_v = \frac{1}{4}\pi U_z^2 \quad [U_z < 0.60]$$

$$= \frac{1}{4}\pi(0.247)^2$$

$$= 0.048$$

The time for a layer to reach a specific consolidation is

$$t = \frac{T_v H^2}{C_v}$$

$$5 \text{ yr} = \frac{0.048 H^2}{C_v}$$

$$\frac{H^2}{C_v} = 104 \text{ yr}$$

For the total settlement of 250 mm,

$$U_z = \frac{\Delta H}{\Delta H_{\text{ult}}} = \frac{250 \text{ mm}}{502 \text{ mm}}$$

$$= 0.498$$

The time factor is

$$T_v = \tfrac{1}{4}\pi(0.498)^2$$

$$= 0.195$$

The time to reach the settlement of 250 mm is

$$t = T_v\left(\frac{H^2}{C_v}\right) = (0.195)(104 \text{ yr})$$

$$= 20.3 \text{ yr}$$

The remaining time to reach a settlement of 250 mm is

$$\Delta t = 20.3 \text{ yr} - 5 \text{ yr}$$

$$= 15.3 \text{ yr} \quad (15 \text{ yr})$$

The answer is (C).

88. Since the backfill is horizontal and the retaining wall is smooth, the coefficient of active earth pressure is

$$k_a = \tan^2\left(45° - \frac{\phi}{2}\right)$$

$$= \tan^2\left(45° - \frac{32°}{2}\right)$$

$$= 0.31$$

To determine the total resultant force and the location of the moment arm, divide the backfill into five components.

For the surcharge at the backfill surface, the effective stress is

$$\sigma'_{a,A} = k_a \sigma'_v = (0.31)(25 \text{ kPa})$$

$$= 7.75 \text{ kPa}$$

For the surcharge and sand above the water table,

$$\sigma'_{a,B} = \sigma'_{a,A} + k_a \gamma H$$

$$= 7.75 \text{ kPa} + (0.31)\left(19.5 \ \frac{\text{kN}}{\text{m}^3}\right)(3.5 \text{ m})$$

$$= 28.9 \text{ kPa}$$

For the surcharge, sand above the water table, and sand below the water table,

$$\sigma'_{a,C} = \sigma'_{a,B} + k_a(\gamma_{\text{sat}} - \gamma_w)H$$

$$= 28.9 \text{ kPa} + (0.31)\left(20.3 \ \frac{\text{kN}}{\text{m}^3} - 9.81 \ \frac{\text{kN}}{\text{m}^3}\right)(6.5 \text{ m})$$

$$= 50.0 \text{ kPa}$$

For the surcharge, sand above the water table, sand below the water table, and water pore pressure,

$$\sigma'_{a,C} + u = \sigma'_{a,C} + \gamma_w H$$

$$= 50.0 \text{ kPa} + \left(9.81 \ \frac{\text{kN}}{\text{m}^3}\right)(6.5 \text{ m})$$

$$= 113.8 \text{ kPa}$$

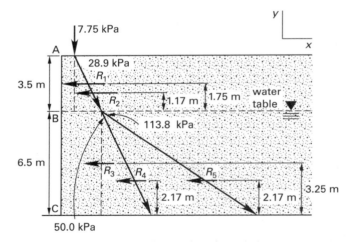

The resultant forces are the areas of the geometrically-shaped pressure distributions.

$$R_1 = (7.75 \text{ kPa})(3.5 \text{ m}) = 27.1 \text{ kN/m}$$

$$R_2 = \left(\tfrac{1}{2}\right)(28.9 \text{ kPa} - 7.75 \text{ kPa})(3.5 \text{ m})$$

$$= 37.0 \text{ kN/m}$$

$$R_3 = (28.9 \text{ kPa})(6.5 \text{ m}) = 187.9 \text{ kN/m}$$

$$R_4 = \left(\tfrac{1}{2}\right)(50.0 \text{ kPa} - 28.9 \text{ kPa})(6.5 \text{ m})$$

$$= 68.6 \text{ kN/m}$$

$$R_5 = \left(\tfrac{1}{2}\right)(113.8 \text{ kPa} - 50.0 \text{ kPa})(6.5 \text{ m})$$

$$= 207.4 \text{ kN/m}$$

Summation of moments about point C gives the location of the resultant active force against the retaining wall.

$$y = \frac{R_1 y_1 + R_2 y_2 + R_3 y_3 + R_4 y_4 + R_5 y_5}{R_1 + R_2 + R_3 + R_4 + R_5}$$

$$= \frac{\begin{array}{c}\left(27.1 \ \frac{kN}{m}\right)(8.25 \ m) + \left(37.0 \ \frac{kN}{m}\right)(7.67 \ m) \\ + \left(187.9 \ \frac{kN}{m}\right)(3.25 \ m) \\ + \left(68.6 \ \frac{kN}{m}\right)(2.17 \ m) \\ + \left(207.4 \ \frac{kN}{m}\right)(2.17 \ m)\end{array}}{\begin{array}{c}27.1 \ \frac{kN}{m} + 37.0 \ \frac{kN}{m} + 187.9 \ \frac{kN}{m} \\ + 68.6 \ \frac{kN}{m} + 207.4 \ \frac{kN}{m}\end{array}}$$

$$= 3.25 \ m \quad (3.3 \ m)$$

The answer is (B).

89. For a loose, natural sand deposit, the coefficient of earth pressure at rest can be estimated as

$$k_0 \approx 1 - \sin\phi = 1 - \sin 29° = 0.52$$

The total at-rest lateral earth pressure is

$$p_o = k_o p_v$$

This can be rewritten as

$$\sigma_h = \sigma_h' + u = k_o \sigma_v' + u = k_o \gamma_b z + u$$
$$= k_o(\gamma_{sat} - \gamma_w) z + \gamma_w z$$
$$= (0.52)\left(19.3 \ \frac{kN}{m^3} - 9.81 \ \frac{kN}{m^3}\right)(10 \ m)$$
$$\quad + \left(9.81 \ \frac{kN}{m^3}\right)(10 \ m)$$
$$= 147.4 \ kN/m^2 \quad (150 \ kPa)$$

The answer is (B).

90. Darcy's law gives the effective (apparent) velocity through the clay liner.

$$v_e = Ki = K\left(\frac{H}{L}\right) = K\left(\frac{h_1 - h_2}{L}\right)$$
$$= \left(2.5 \times 10^{-7} \ \frac{mm}{s}\right)\left(\frac{2.25 \ m - 0 \ m}{1.00 \ m}\right)$$
$$= 5.63 \times 10^{-7} \ mm/s$$

The problem asks for pore velocity, v_{pore}. Porosity, n, relates the pore velocity to the effective velocity of the flow.

$$e = \frac{n}{1 - n}$$
$$n = \frac{e}{1 + e} = \frac{0.88}{1 + 0.88}$$
$$= 0.47$$
$$v_{pore} = \frac{v_e}{n}$$
$$= \frac{5.63 \times 10^{-7} \ \frac{mm}{s}}{0.47}$$
$$= 1.2 \times 10^{-6} \ mm/s$$

The answer is (C).

91. Since the backfill is horizontal and the retaining wall is smooth, the coefficient of active earth pressure is

$$k_a = \tan^2\left(45° - \frac{\phi}{2}\right)$$
$$= \tan^2\left(45° - \frac{30°}{2}\right)$$
$$= 0.33$$

The active force per unit length can be calculated using Rankine theory.

$$R_a = \tfrac{1}{2} k_a \gamma H^2$$
$$= \left(\frac{1}{2}\right)(0.33)\left(20 \ \frac{kN}{m^3}\right)(2.2 \ m)^2$$
$$= 16.0 \ kN/m$$

The overturning moment per unit length of wall is given by the following equation. M is the overturning moment per unit length of wall, and M_B is the total overturning moment.

$$M = \frac{M_B}{L} = R_a\left(\frac{H}{3}\right)$$
$$= \left(16.0 \ \frac{kN}{m}\right)\left(\frac{2.2 \ m}{3}\right)$$
$$= 11.7 \ kN \cdot m/m$$

The weight of the wall (per unit length of wall) is the only vertical force.

$$W = \frac{P}{L} = HB\gamma_{concrete}$$
$$= (2.8 \ m)(1.0 \ m)\left(25 \ \frac{kN}{m^3}\right)$$
$$= 70 \ kN/m$$

The vertical pressure at point A is the maximum pressure beneath the wall.

$$p_A = p_{max} = \frac{P}{BL}\left(1 + \frac{6\epsilon}{B}\right)$$
$$= \frac{W}{B}\left(1 + \frac{6\epsilon}{B}\right)$$

The eccentricity is

$$\epsilon = \frac{M_B}{P} = \frac{M}{W}$$
$$= \frac{11.7 \ \dfrac{kN \cdot m}{m}}{70 \ \dfrac{kN}{m}}$$
$$= 0.167 \ m$$

$$p_A = \frac{W}{B}\left(1 + \frac{6\epsilon}{B}\right)$$
$$= \left(\frac{70 \ \dfrac{kN}{m}}{1.0 \ m}\right)\left(1 + \frac{(6)(0.167 \ m)}{1.0 \ m}\right)$$
$$= 140 \ kN/m^2 \quad (140 \ kPa)$$

The answer is (D).

92. This problem is solved using the standard weight-volume relationships for soils.

The moisture content is

$$w = \frac{W_w}{W_s} = \frac{m_w}{m_s}$$
$$= \frac{1733 \ g - 1287 \ g}{1287 \ g} \times 100\%$$
$$= 34.7\%$$

Since the soil sample is saturated, the degree of saturation is 100%. The void ratio is

$$Se = wSG$$
$$e = \frac{wSG}{S} = \frac{(34.7\%)(2.7)}{100\%}$$
$$= 0.937$$

The total unit weight is

$$\gamma = \frac{(SG + Se)\gamma_w}{1 + e}$$
$$= \frac{(2.7 + 0.937)\left(9.81 \ \dfrac{kN}{m^3}\right)}{1 + 0.937}$$
$$= 18.4 \ kN/m^3$$

The answer is (D).

93. For uniform sands, the permeability can be estimated using the following equation.

$$K_{cm/s} \approx CD_{10,mm}^2$$

For medium sand, well graded, the coefficient C varies between 0.8 and 1.2. The effective grain size for this soil is approximately

$$D_{10,mm} = 0.15 \ mm$$

Assuming C equals 1.0,

$$K_{cm/s} \approx (1.0)(0.15)^2$$
$$\approx 0.0225 \ cm/s \quad (2 \times 10^{-2} \ cm/s)$$

The answer is (B).

94. From the elevation information given, the total head is

$$h = 365 \ m - 360 \ m = 5 \ m$$

From the flow net, the total number of head drops, N_D, is 14, and the number of head drops to point A is 12. The depth at point A is

$$z_A = 360 \ m - 350 \ m$$
$$= 10 \ m$$

The head lost up to point A is

$$\Delta h_A = h\left(\frac{\text{number of head drops to A}}{N_D}\right)$$
$$= (5 \ m)\left(\frac{12}{14}\right)$$
$$= 4.29 \ m$$

The pore pressure at point A is

$$p_A = \gamma_w(h + z_A - \Delta h_A)$$
$$= \left(9.81 \ \frac{kN}{m^3}\right)(5 \ m + 10 \ m - 4.29 \ m)$$
$$= 105 \ kN/m^2 \quad (105 \ kPa)$$

The answer is (B).

95. The different vertical coefficients of permeability in these stratified anisotropic soils can be combined into one effective vertical coefficient of permeability, K.

$$K = \frac{\sum H_j}{\sum \dfrac{H_j}{K_j}}$$

$$= \frac{2.5 \text{ m} + 1.0 \text{ m} + 3.0 \text{ m}}{\dfrac{2.5 \text{ m}}{3.0 \times 10^{-5} \dfrac{\text{mm}}{\text{s}}} + \dfrac{1.0 \text{ m}}{2.0 \times 10^{-6} \dfrac{\text{mm}}{\text{s}}}}$$
$$+ \dfrac{3.0 \text{ m}}{3.8 \times 10^{-6} \dfrac{\text{mm}}{\text{s}}}$$

$$= 4.7 \times 10^{-6} \text{ mm/s}$$

From Darcy's Law,

$$V = Qt = KiAt$$
$$= K\frac{H}{L}At = K\left(\frac{h_1 - h_2}{L}\right)At$$

$$= \frac{(4.7 \times 10^{-6} \text{ mm/s})}{}$$
$$\times \left(\frac{1.5 \text{ m} - 0 \text{ m}}{2.5 \text{ m} + 1.0 \text{ m} + 3.0 \text{ m}}\right)$$
$$\times (5000 \text{ m}^2)(6 \text{ mo})\left(30\frac{\text{d}}{\text{mo}}\right)$$
$$= \frac{\times \left(24\dfrac{\text{h}}{\text{d}}\right)\left(3600\dfrac{\text{s}}{\text{h}}\right)}{1000 \dfrac{\text{mm}}{\text{m}}}$$

$$= 84.3 \text{ m}^3 \quad (85 \text{ m}^3)$$

The answer is (A).

96. The effective area is the greatest possible portion of the footing such that the resultant force passes through its centroid. The eccentricities are

$$\epsilon_B = \frac{0.9 \text{ m} + 0.6 \text{ m}}{2} - 0.6 \text{ m} = 0.15 \text{ m}$$

$$\epsilon_L = \frac{1 \text{ m} + 2 \text{ m}}{2} - 1 \text{ m} = 0.5 \text{ m}$$

Given a rectangular footing of width B and length L, and a load with eccentricity ϵ of 0.15 m in the B-direction and 0.5 m in the L-direction, the equivalent width B' and length L' of the effective area are

$$B' = B - 2\epsilon_B$$
$$= 1.5 \text{ m} - (2)(0.15 \text{ m})$$
$$= 1.2 \text{ m}$$
$$L' = L - 2\epsilon_L$$
$$= 3.0 \text{ m} - (2)(0.5 \text{ m})$$
$$= 2.0 \text{ m}$$

The effective area is

$$A_e = L'B'$$
$$= (2.0 \text{ m})(1.2 \text{ m})$$
$$= 2.4 \text{ m}^2$$

The answer is (C).

97. The ultimate bearing capacity is given by the following equation.

$$q_{\text{ult}} = c\lambda_{cs}\lambda_{cd}N_c + q\lambda_{qs}\lambda_{qd}N_q + \tfrac{1}{2}\lambda_{\gamma s}\lambda_{\gamma d}\gamma B N_\gamma$$

From a table of Terzaghi bearing capacity factors, when ϕ equals $0°$, N_c is 5.7, N_q is 1.0, and N_γ is 0.

The shape and depth factors are given as

$$\lambda_{qs} = \lambda_{\gamma s} = 1$$
$$\lambda_{qd} = \lambda_{\gamma d} = 1$$

The value of B/L is so small it can be taken as zero, so

$$\lambda_{cs} = 1 + 0.2\left(\frac{B}{L}\right)\tan^2\left(45° + \frac{\phi}{2}\right)$$
$$= 1$$
$$\lambda_{cd} = 1 + 0.2\left(\frac{D_f}{B}\right)\tan\left(45° + \frac{\phi}{2}\right)$$
$$= 1 + (0.2)\left(\frac{1.0 \text{ m}}{2.0 \text{ m}}\right)\tan\left(45° + \frac{0°}{2}\right)$$
$$= 1.1$$

The surcharge is

$$q = (\gamma_{\text{sat}} - \gamma_w)D_f$$
$$= \left(18.5 \frac{\text{kN}}{\text{m}^3} - 9.8 \frac{\text{kN}}{\text{m}^3}\right)(1.0 \text{ m})$$
$$= 8.7 \text{ kN/m}^2$$

Therefore,

$$q_{\text{ult}} = c\lambda_{cd}N_c + q$$
$$= \left(110 \frac{\text{kN}}{\text{m}^2}\right)(1.1)(5.7) + 8.7 \frac{\text{kN}}{\text{m}^2}$$
$$= 698.4 \text{ kN/m}^2$$
$$P_{\text{ult}} = q_{\text{ult}}B$$
$$= \left(698.4 \frac{\text{kN}}{\text{m}^2}\right)(2.0 \text{ m})$$
$$= 1397 \text{ kN/m} \quad (1400 \text{ kN/m})$$

The answer is (D).

98. The rock quality designation, RQD, is defined as the total length of all intact pieces 10 cm or longer, divided by the total length of the core, expressed as a percentage.

$$\text{RQD} = \left(\frac{89 \text{ cm}}{123 \text{ cm}}\right) \times 100\%$$
$$= 72\%$$

The answer is (C).

99. The ultimate pullout capacity of the pile is composed of the pile weight and the skin friction. The weight of the pile is

$$W = LA\gamma_c = L\left(\frac{\pi d^2}{4}\right)\gamma_c$$
$$= (6.0 \text{ m})\left(\frac{\pi(0.30 \text{ m})^2}{4}\right)\left(25 \frac{\text{kN}}{\text{m}^3}\right)$$
$$= 10.6 \text{ kN}$$

The skin friction area is πdL. The skin friction is

$$Q_f = k_h \sigma_v' \tan \delta \pi dL$$
$$= k_h \gamma \left(\frac{L}{2}\right) \tan \delta \pi dL$$
$$= (1.1)\left(20 \frac{\text{kN}}{\text{m}^3}\right)\left(\frac{6.0 \text{ m}}{2}\right)(\tan 25°)$$
$$\times \pi(0.30 \text{ m})(6.0 \text{ m})$$
$$= 174.0 \text{ kN}$$

σ_v' is the average effective vertical pressure of the soil along the pipe and is equal to $\gamma L/2$.

The pullout capacity is

$$P = W + Q_f = 10.6 \text{ kN} + 174.0 \text{ kN}$$
$$= 184.6 \text{ kN} \quad (180 \text{ kN})$$

The answer is (C).

100. The total primary consolidation settlement is

$$S_p = \frac{H_o C_c \log\left(\frac{p_o' + \Delta p_v'}{p_o'}\right)}{1 + e_o}$$
$$= \frac{(8 \text{ m})(0.26)\log\left(\frac{240 \text{ kPa} + 130 \text{ kPa}}{240 \text{ kPa}}\right)}{1 + 1.02}$$
$$= 0.194 \text{ m} \quad (190 \text{ mm})$$

The answer is (C).

101. The time factor for the clay layer is

$$T_v = \frac{C_v t}{H_d^2}$$
$$= \frac{\left(7.6 \times 10^{-8} \frac{\text{m}^2}{\text{s}}\right)(3 \text{ yr})\left(365 \frac{\text{d}}{\text{yr}}\right)}{\left(\frac{6.0 \text{ m}}{2}\right)^2}$$
$$= 0.80$$

Interpolating from a table of time factors, the degree of consolidation, U_z, for a time factor of 0.80 is about 88%, or most nearly 90%.

The answer is (D).

102. The time factor, T_v, for a specified average degree of consolidation, U_z, is found from a table of approximate time factors.

For an average degree of consolidation of 0.90, T_v is approximately equal to 0.85.

The time required to achieve this degree of consolidation is

$$t = \frac{T_v H_d^2}{C_v}$$
$$= \frac{(0.85)\left(\frac{20 \text{ m}}{2}\right)^2}{\left(4.3 \times 10^{-7} \frac{\text{m}^2}{\text{s}}\right)\left(3600 \frac{\text{s}}{\text{h}}\right)\left(24 \frac{\text{h}}{\text{d}}\right)\left(365 \frac{\text{d}}{\text{yr}}\right)}$$
$$= 6.3 \text{ yr}$$

The answer is (C).

103. For one-dimensional loading, the excess pore water pressure at beginning of loading, u_i, is 100 kPa. As time passes, the excess pore water pressure decreases.

The answer is (D).

104. Two formulas for calculating dry unit weight are

$$\gamma_d = \frac{\gamma}{1 + w}$$
$$\gamma_d = \frac{\text{SG}\gamma_w}{1 + e}$$

Therefore,

$$\frac{\gamma}{1+w} = \frac{SG\gamma_w}{1+e}$$

$$e = \frac{SG(1+w)\gamma_w}{\gamma} - 1$$

$$= \frac{(2.72)(1+0.08)\left(9.81 \ \frac{kN}{m^3}\right)}{17.6 \ \frac{kN}{m^3}} - 1$$

$$= 0.64$$

The degree of saturation is

$$S = \frac{SGw}{e} = \frac{(2.72)(0.08)}{0.64} \times 100\%$$

$$= 34\%$$

The answer is (D).

105. In order to determine the appropriate active soil pressure envelope to use, it is necessary to determine if the clay is soft, medium, or stiff by calculating the stability number, N_o.

$$N_o = \frac{\gamma H}{c} = \frac{\left(18.3 \ \frac{kN}{m^3}\right)(8 \ m)}{23 \ \frac{kN}{m^2}}$$

$$= 6.4$$

N_o is greater than 6, so the braced cut is in soft clay. The pressure distribution for soft clay is shown.

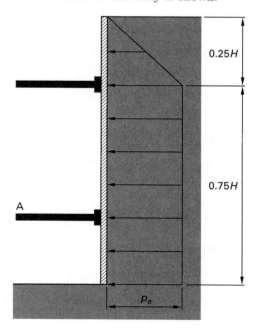

In this pressure distribution (which is strictly applicable for $H \geq 6$ m and a water table below the bottom of the cut), active pressure is

$$p_a = \gamma H - 4c$$

$$= \left(18.3 \ \frac{kN}{m^3}\right)(8 \ m) - (4)\left(23 \ \frac{kN}{m^2}\right)$$

$$= 54.4 \ kN/m^2 \quad (54.4 \ kPa)$$

The pressure distribution is

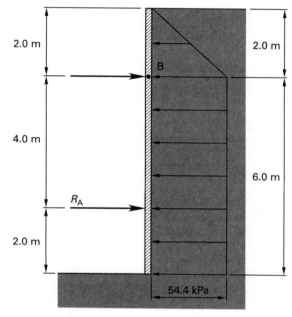

By resolving the pressure distribution into resultant forces and taking moments about point B, the force per unit width is

$$\sum M_B = R_A(4.0 \ m) - \left(54.4 \ \frac{kN}{m^2}\right)(4.0 \ m + 2.0 \ m)$$

$$\times \left(\frac{4.0 \ m + 2.0 \ m}{2}\right) + \left(\frac{1}{2}\right)\left(54.4 \ \frac{kN}{m^2}\right)$$

$$\times (2.0 \ m)\left(\frac{2.0 \ m}{3}\right)$$

$$= 0$$

$$R_A = \frac{\left(54.4 \ \frac{kN}{m^2}\right)(6.0 \ m)\left(\frac{6.0 \ m}{2}\right)}{4.0 \ m}$$
$$- \left(\frac{1}{2}\right)\left(54.4 \ \frac{kN}{m^2}\right)(2.0 \ m)\left(\frac{2.0 \ m}{3}\right)$$
over $4.0 \ m$

$$= 235.7 \ kN \quad \text{[per meter of width]}$$

Since the struts are spaced 4 m apart, the strut force is

$$(4)(235.7 \ kN) = 942.9 \ kN \quad (940 \ kN)$$

The answer is (C).

106. From the grain size distribution, D_{10} is 0.009 mm, D_{30} is 0.45 mm, and D_{60} is 8.5 mm.

The coefficient of curvature is

$$C_z = \frac{D_{30}^2}{D_{10}D_{60}}$$

$$= \frac{(0.45 \text{ mm})^2}{(0.009 \text{ mm})(8.5 \text{ mm})}$$

$$= 2.6$$

The answer is (A).

107. The unit skin friction along the pile is found by the equation

$$f_s = k\sigma'_v \tan \delta$$

The critical depth is $15B$. For a depth, z, from 0 to $15B$,

$$\sigma'_v = \gamma z$$

$$= \left(18.8 \ \frac{\text{kN}}{\text{m}^3}\right)z$$

For a depth greater or equal to $15B$,

$$\sigma'_v = \gamma z = \gamma 15B$$

$$= \left(18.8 \ \frac{\text{kN}}{\text{m}^3}\right)(15)(0.254 \text{ m})$$

$$= 71.6 \text{ kN/m}^2 \quad (71.6 \text{ kPa})$$

The resulting vertical effective stress distribution is shown.

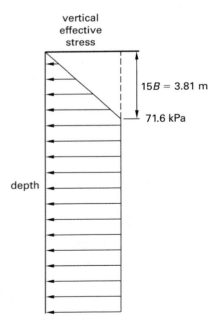

For the four sides of the square pile ($4BL'$) and the average pressure, the skin-friction capacity along the pile from depth 0 to $15B$ is

$$Q_f = 4BL'f_{\text{ave}} = 4BL'\tfrac{1}{2}k\sigma'_v \tan \delta$$

$$= (4)(0.254 \text{ m})(15)(0.254 \text{ m})\left(\tfrac{1}{2}\right)$$

$$\times (1.6)\left(71.6 \ \frac{\text{kN}}{\text{m}^2}\right)(\tan 0.6 \times 35°)$$

$$= 85.1 \text{ kN}$$

The skin-friction capacity along the pile from depth $15B$ to 10 m is

$$Q_f = 4B(L - L')f_{z=15D}$$

$$= (4)(0.254 \text{ m})(10 \text{ m} - 3.81 \text{ m})$$

$$\times (1.6)\left(71.6 \ \frac{\text{kN}}{\text{m}^2}\right)\tan(0.6 \times 35°)$$

$$= 276.6 \text{ kN}$$

The total frictional resistance along the pile is

$$Q_{s,\text{total}} = \sum Q_s = 85.1 \text{ kN} + 276.6 \text{ kN}$$

$$= 362 \text{ kN} \quad (360 \text{ kN})$$

The answer is (C).

108. Rankine theory for a passive earth condition should be used to solve this problem. The passive earth coefficient is

$$k_p = \tan^2\left(45° + \frac{\phi}{2}\right)$$

$$= \tan^2\left(45° + \frac{28°}{2}\right)$$

$$= 2.77$$

The total passive force per meter of wall is

$$R = p_{h,\text{ave}}H$$

$$= \tfrac{1}{2}\gamma H^2 k_p + 2c\sqrt{k_p}H$$

$$= \begin{array}{l} \left(\tfrac{1}{2}\right)\left(18.8 \ \frac{\text{kN}}{\text{m}^3}\right)(3.0 \text{ m})^2(2.77) \\ + (2)\left(16 \ \frac{\text{kN}}{\text{m}^2}\right)\sqrt{2.77}(3.0 \text{ m}) \end{array}$$

$$= 344.1 \text{ kN/m} \quad (400 \text{ kN/m})$$

The answer is (B).

109. The effective minor stress at failure is

$$\bar{\sigma}_3 = 200 \text{ kPa}$$

The effective major stress at failure is

$$\bar{\sigma}_1 = 200 \text{ kPa} + 468 \text{ kPa} = 668 \text{ kPa}$$

The angle of internal friction can be calculated using the following equation.

$$\frac{\bar{\sigma}_1}{\bar{\sigma}_3} = \frac{1 + \sin\phi}{1 - \sin\phi}$$

$$\sin\phi = \frac{\bar{\sigma}_1 - \bar{\sigma}_3}{\bar{\sigma}_1 + \bar{\sigma}_3}$$

$$= \frac{668 \text{ kPa} - 200 \text{ kPa}}{668 \text{ kPa} + 200 \text{ kPa}}$$

$$= 0.539$$

$$\phi = 32.6° \quad (33°)$$

The answer is (C).

110. The buoyant unit weight is

$$\gamma_b = \gamma_{\text{sat}} - \gamma_w$$

$$= \left(\frac{\text{SG} + e}{1 + e}\right)\gamma_w - \gamma_w$$

$$= \left(\frac{2.66 + 0.62}{1 + 0.62}\right)\left(9.81 \ \frac{\text{kN}}{\text{m}^3}\right) - 9.81 \ \frac{\text{kN}}{\text{m}^3}$$

$$= 10.1 \text{ kN/m}^3 \quad (10 \text{ kN/m}^3)$$

The answer is (A).

111. The ultimate bearing capacity is given by the equation

$$q_{\text{ult}} = \tfrac{1}{2}\gamma B N_\gamma + c N_c + (p_q + \gamma D_f)N_q$$

The footing is placed near the ground surface, and p_q equals 0 kN/m², so $p_q + \gamma D_f$ is zero.

$$q_{\text{ult}} = \tfrac{1}{2}\gamma B N_\gamma + c N_c$$

A table of Terzaghi bearing capacity factors shows that when ϕ equals 25°, N_c equals 25.1, and N_γ equals 9.7.

$$q_{\text{ult}} = \tfrac{1}{2}\gamma B N_\gamma + c N_c$$

$$= \left(\frac{1}{2}\right)\left(18.6 \ \frac{\text{kN}}{\text{m}^3}\right)(1.5 \text{ m})(9.7)$$

$$+ \left(14 \ \frac{\text{kN}}{\text{m}^2}\right)(25.1)$$

$$= 486.7 \text{ kN/m}^2$$

$$q_{\text{actual}} = \frac{P_{\text{actual}}}{B} = \frac{596 \ \frac{\text{kN}}{\text{m}}}{1.5 \text{ m}}$$

$$= 397.3 \text{ kN/m}^2$$

The factor of safety is

$$F = \frac{q_{\text{ult}}}{q_{\text{actual}}} = \frac{486.7 \ \frac{\text{kN}}{\text{m}^2}}{397.3 \ \frac{\text{kN}}{\text{m}^2}}$$

$$= 1.22 \quad (1.2)$$

The answer is (B).

112. Coulomb's equation relates soil strength to the normal stress on the failure plane.

$$S = \tau = c + \sigma \tan\phi$$

The cohesiveness of sand, c, is 0, so

$$\tau = \sigma \tan\phi$$

$$\tan\phi = \frac{\tau}{\sigma} = \frac{63.4 \ \frac{\text{kN}}{\text{m}^2}}{100 \ \frac{\text{kN}}{\text{m}^2}}$$

$$= 0.634$$

Use the same relationship to calculate the shear strength of that soil for a normal stress of 75 kN/m².

$$\tau = \sigma \tan\phi = \left(75 \ \frac{\text{kN}}{\text{m}^2}\right)(0.634)$$

$$= 47.6 \text{ kN/m}^2$$

The shear force required to cause failure is

$$S = \tau A = \left(47.6 \ \frac{\text{kN}}{\text{m}^2}\right)(0.06 \text{ m})^2$$

$$= 0.17 \text{ kN}$$

The answer is (A).

113. During an undrained test, the volume of the sample does not change.

$$V = HA = H_0 \frac{\pi d_0^2}{4}$$

$$= H_f \frac{\pi d_f^2}{4}$$

$$d_f = \sqrt{\frac{H_0}{H_f}} d_0$$

$$= \sqrt{\frac{9.1 \text{ cm}}{8.67 \text{ cm}}}(4.0 \text{ cm})$$

$$= 4.1 \text{ cm} \quad (0.041 \text{ m})$$

The maximum principal stress is

$$\sigma_1 = \frac{P}{A_f} = \frac{P}{\frac{\pi d_f^2}{4}}$$

$$= \frac{0.43 \text{ kN}}{\frac{\pi (0.041 \text{ m})^2}{4}}$$

$$= 326 \text{ kN/m}^2$$

The undrained shear strength is

$$S_u = c = \frac{\sigma_1}{2}$$

$$= \frac{326 \frac{\text{kN}}{\text{m}^2}}{2}$$

$$= 163 \text{ kN/m}^2 \quad (160 \text{ kPa})$$

The answer is (B).

114. Since this test is performed under drained conditions, there is no pore water pressure.

$$\sigma_1' = \sigma_3' + \Delta\sigma_{D,f}$$

$$= 280 \frac{\text{kN}}{\text{m}^2} + 410 \frac{\text{kN}}{\text{m}^2}$$

$$= 690 \text{ kN/m}^2$$

The clay has no drained cohesion, so the effective principal stresses at failure can be related by this equation, solving for ϕ.

$$\frac{\sigma_1'}{\sigma_3'} = \frac{1 + \sin\phi}{1 - \sin\phi}$$

The following equation is also commonly used.

$$\sigma_1' = \sigma_3' \tan^2\left(45° + \frac{\phi}{2}\right)$$

Solving for ϕ,

$$\phi = 2\arctan\sqrt{\frac{\sigma_1'}{\sigma_3'}} - 45°$$

$$= (2)\left(\arctan\sqrt{\frac{690 \frac{\text{kN}}{\text{m}^2}}{280 \frac{\text{kN}}{\text{m}^2}}} - 45°\right)$$

$$= 25°$$

The angle between the failure plane and the major principal plane is

$$\alpha = 45° + \frac{\phi}{2}$$

$$= 45° + \frac{25°}{2}$$

$$= 57.5°$$

The shear stress on the failure plane is

$$\tau_f = \frac{\sigma_1' - \sigma_3'}{2} \sin 2\theta$$

$$= \left(\frac{690 \frac{\text{kN}}{\text{m}^2} - 280 \frac{\text{kN}}{\text{m}^2}}{2}\right) \sin\left((2)(57.5°)\right)$$

$$= 185.8 \text{ kN/m}^2 \quad (190 \text{ kPa})$$

The answer is (A).

115. The void ratio of the sand can be calculated using the following formula.

$$e = \frac{SG\gamma_w}{\gamma_d} - 1$$

$$= \frac{(2.65)\left(9.81 \frac{\text{kN}}{\text{m}^3}\right)}{16.5 \frac{\text{kN}}{\text{m}^3}} - 1$$

$$= 0.576$$

Relative density can be calculated with the following formula.

$$D_r = \frac{e_{\max} - e}{e_{\max} - e_{\min}}$$

$$= \frac{0.78 - 0.576}{0.78 - 0.41}$$

$$= 0.55$$

The answer is (B).

116. The group index is computed as

$$I_g = (F_{200} - 35)(0.2 + 0.005(LL - 40))$$
$$+ (0.01)(F_{200} - 15)(PI - 10)$$
$$= (45 - 35)(0.2 + (0.005)(40 - 40))$$
$$+ (0.01)(45 - 15)(13 - 10)$$
$$= 2.9 \quad (3)$$

The percentage passing a no. 200 sieve is greater than 36%, the liquid limit is 40, and the plasticity index is greater than 11. The AASHTO classification and group index of this soil is A-6 (3).

The answer is (C).

117. Use a USCS table to solve this problem.

As less than 50% passes a no. 200 sieve, the soil is coarse grained.

More than half the coarse fraction is finer than no. 4, so the soil is a sand (S).

More than 12% passes a no. 200 sieve, so the soil will be classified as either SM or SC.

The plasticity index is

$$PI = LL - PL = 55 - 20 = 35$$

With a liquid limit of 55 and a plasticity index of 35, the fine-grained fraction of the soil classifies as highly plastic clay (CH).

The soil is therefore classified as SC.

The answer is (A).

118. The active force can be calculated using the Rankine theory.

$$R_a = \tfrac{1}{2}k_a\gamma H^2 = \left(\frac{1}{2}\right)(0.33)\left(20\ \frac{\text{kN}}{\text{m}^3}\right)(5.5\ \text{m})^2$$
$$= 99.8\ \text{kN/m}$$

The backfill is horizontal, and the wall face is vertical.

$$k_a = \tan^2\left(45° - \frac{\phi}{2}\right)$$
$$= \tan^2\left(45° - \frac{30°}{2}\right)$$
$$= 0.33$$

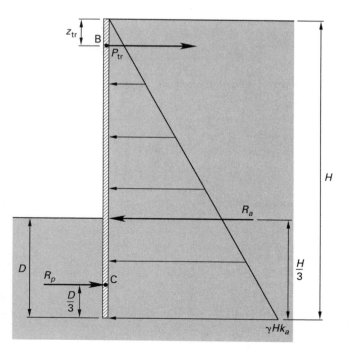

All forces are in equilibrium, so the sum of the moments about point B is zero. Taking moments about B allows the tie rod force to be ignored. The moment balance is

$$\sum M_{\text{B}} = R_a\left(\tfrac{2}{3}H - z_{\text{tr}}\right) - R_p\left(H - z_{\text{tr}} - \tfrac{1}{3}D\right)$$
$$= 0$$
$$R_p = R_a\left(\frac{\tfrac{2}{3}H - z_{\text{tr}}}{H - z_{\text{tr}} - \tfrac{1}{3}D}\right)$$
$$= \left(99.8\ \frac{\text{kN}}{\text{m}}\right)\left(\frac{\left(\frac{2}{3}\right)(5.5\ \text{m}) - 0.5\ \text{m}}{5.5\ \text{m} - 0.5\ \text{m} - \left(\frac{1}{3}\right)(2.1\ \text{m})}\right)$$
$$= 73.5\ \text{kN/m} \quad (70\ \text{kN/m})$$

The answer is (A).

119. To have a fully compensated foundation, the total weight of the excavated soil must be equal to the total load on the foundation.

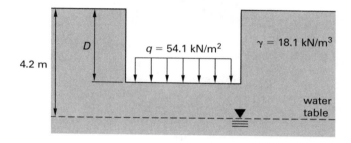

The pressure due to the total dead and live load is

$$q = \frac{P}{BL}$$
$$= \frac{33\,540\ \text{kN}}{(20\ \text{m})(31\ \text{m})}$$
$$= 54.1\ \text{kN/m}^2$$

The embedment depth required to have a fully compensated foundation is

$$\gamma D = q$$
$$D = \frac{q}{\gamma}$$
$$= \frac{54.1\ \dfrac{\text{kN}}{\text{m}^2}}{18.1\ \dfrac{\text{kN}}{\text{m}^3}}$$
$$= 3.0\ \text{m}$$

The answer is (C).

120. The factor of safety against sliding can be expressed as

$$\text{FS}_{sl} = \frac{\sum R_r}{\sum R_o}$$

$\sum R_r$ is the sum of forces per unit length of wall resisting sliding, and $\sum R_o$ is the sum of forces per unit length of wall causing sliding.

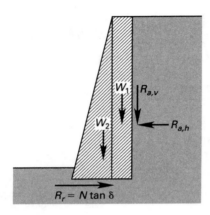

The active force is

$$R_a = \tfrac{1}{2}k_a\gamma H^2$$
$$= \left(\tfrac{1}{2}\right)(0.31)\left(20.2 \ \frac{\text{kN}}{\text{m}^3}\right)(4.3 \ \text{m})^2$$
$$= 57.9 \ \text{kN/m}$$

The horizontal and vertical components of the active force are

$$R_{a,h} = R_a \cos\delta$$
$$= \left(57.9 \ \frac{\text{kN}}{\text{m}}\right)\cos 20°$$
$$= 54.4 \ \text{kN/m}$$

$$R_{a,v} = R_a \sin\delta$$
$$= \left(57.9 \ \frac{\text{kN}}{\text{m}}\right)\sin 20°$$
$$= 19.8 \ \text{kN/m}$$

The weight of the wall can be subdivided into W_1 and W_2.

$$W_1 = H b_1 \gamma_{\text{concrete}}$$
$$= (4.3 \ \text{m})(0.4 \ \text{m})\left(25 \ \frac{\text{kN}}{\text{m}^3}\right)$$
$$= 43.0 \ \text{kN/m}$$
$$W_2 = \tfrac{1}{2}H b_2 \gamma_{\text{concrete}}$$
$$= \left(\tfrac{1}{2}\right)(4.3 \ \text{m})(2.1 \ \text{m})\left(25 \ \frac{\text{kN}}{\text{m}^3}\right)$$
$$= 112.9 \ \text{kN/m}$$
$$W = W_1 + W_2$$
$$= 43.0 \ \frac{\text{kN}}{\text{m}} + 112.9 \ \frac{\text{kN}}{\text{m}}$$
$$= 155.9 \ \text{kN/m}$$

The total normal force acting on the base of the wall is

$$N = W + R_{a,v}$$
$$= 155.9 \ \frac{\text{kN}}{\text{m}} + 19.8 \ \frac{\text{kN}}{\text{m}}$$
$$= 175.7 \ \text{kN/m}$$

The only force that resists sliding per unit length of wall is

$$R_r = N \tan\delta$$
$$= \left(175.7 \ \frac{\text{kN}}{\text{m}}\right)\tan 20°$$
$$= 63.9 \ \text{kN/m}$$

The horizontal component of the active force is the only force that tends to slide the wall.

$$R_o = P_{a,h}$$
$$= 54.4 \ \text{kN/m}$$

The factor of safety against sliding is

$$\text{FS}_{sl} = \frac{R_r}{R_o} = \frac{63.9 \ \frac{\text{kN}}{\text{m}}}{54.4 \ \frac{\text{kN}}{\text{m}}}$$
$$= 1.17 \quad (1.2)$$

The answer is (B).

Solutions
Afternoon Session
Structural

121. The HL-93 consists of an HS20-44 design truck combined with a 640 lbf/ft lane load. The resultant of the three wheel loads for an HS20-44 loading is a 72 kips force located 4.67 ft from the 32 kips center force.

$$\bar{x} = \frac{\sum P_i x_i}{R} = \frac{(32 \text{ kips})(14 \text{ ft}) + (8 \text{ kips})(28 \text{ ft})}{72 \text{ kips}}$$
$$= 9.33 \text{ ft}$$

Maximum wheel-load bending moment occurs when the midspan lies halfway between the resultant and the central 32 kips force. Thus, the position for maximum wheel-load bending moment is

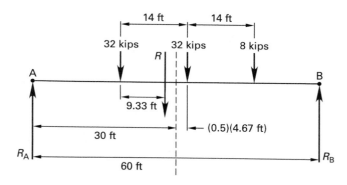

Maximum wheel-load bending moment from the truck occurs under the 32 kips load to the right of midspan.

$$R_A = \frac{\sum rF}{L} = \frac{(72 \text{ kips})\left(30 \text{ ft} + (0.5)(4.67 \text{ ft})\right)}{60 \text{ ft}}$$
$$= 38.8 \text{ kips}$$

$$M_{\text{truck}} = \sum rF = (38.8 \text{ kips})\left(30 \text{ ft} + (0.5)(4.67 \text{ ft})\right)$$
$$- (32 \text{ kips})(14 \text{ ft})$$
$$= 807 \text{ ft-kips}$$

The maximum moment due to the lane loading is

$$M_{\text{LL}} = \frac{wL^2}{8} = \frac{\left(0.640 \dfrac{\text{kip}}{\text{ft}}\right)(60 \text{ ft})^2}{8}$$
$$= 288 \text{ ft-kips}$$

The AASHTO specification requires an increase in the wheel-load bending moment to account for dynamic loading, but this is not applied to the dead or lane loading. A multiple presence factor, MPF, is required when only one lane is loaded.

$$\text{IM (dynamic load allowance)} = 0.33$$
$$\text{MPF (multiple presence factor)} = 1.2$$
$$\text{DC (component dead load factor)} = 1.25$$
$$\text{LL (live load factor)} = 1.75$$

$$M_u = (\text{DC})M_D + (\text{LL})(\text{MPF})$$
$$\times \left(M_{\text{LL}} + (1 + \text{IM})M_{\text{truck}}\right)$$
$$= (1.25)(500 \text{ ft-kips}) + (1.75)(1.2)$$
$$\times \left(288 \text{ ft-kips} + (1 + 0.33)(807 \text{ ft-kips})\right)$$
$$= 3483.8 \text{ ft-kips} \quad (3500 \text{ ft-kips})$$

The answer is (C).

122. The resultant lateral force is

$$V = wL = \left(0.4 \frac{\text{kip}}{\text{ft}}\right)(160 \text{ ft})$$
$$= 64 \text{ kips}$$

This resultant force acts 80 ft from the west wall. The center of rigidity of the wall group is

$$\bar{x} = \frac{\sum R_i x_i}{\sum R_i} = \frac{4R(0 \text{ ft}) + 3R(120 \text{ ft}) + 3R(160 \text{ ft})}{4R + 3R + 3R}$$
$$= 84 \text{ ft} \quad \text{[from the west side of wall A]}$$

(This disregards the accidental torsion of 5% that may be required by code or ASCE/SEI7.)

From symmetry,

$$\bar{y} = 30 \text{ ft} \quad \text{[from the south wall]}$$

The wall system is subjected to a torsional moment of

$$M_t = V\left(\bar{x} - \frac{L}{2}\right) = (64 \text{ kips})\left(84 \text{ ft} - \frac{160 \text{ ft}}{2}\right)$$
$$= 256 \text{ ft-kips} \quad \text{[clockwise]}$$

The polar moment of inertia for the walls resisting the torsional moment is

$$
\begin{aligned}
J &= \sum (R_{yi}x_i^2 + R_{xi}y_i^2) \\
&= 4R(-84 \text{ ft})^2 + 3R(120 \text{ ft} - 84 \text{ ft})^2 \\
&\quad + 3R(160 \text{ ft} - 84 \text{ ft})^2 \\
&\quad + R(30 \text{ ft})^2 + R(30 \text{ ft})^2 \\
&= 51{,}240R \text{ ft}^2
\end{aligned}
$$

The maximum lateral force resisted by wall A is the combined direct force plus the force caused by the torsional moment, both acting in the same sense.

$$
\begin{aligned}
V_A &= \frac{4R}{\sum R_{yi}} V + \frac{M_t R_i x_i}{J} \\
&= \left(\frac{4R}{10R}\right)(64 \text{ kips}) + \frac{(256 \text{ ft-kips})4R(84 \text{ ft})}{51{,}240R \text{ ft}^2} \\
&= 27.3 \text{ kips} \quad (27 \text{ kips})
\end{aligned}
$$

The answer is (C).

123. The plywood diaphragm is considered flexible, and the lateral forces transfer to the shear wall on the basis of their tributary width. Thus, the lateral force acting on the shear wall at line 2 is

$$
\begin{aligned}
V &= \sum wB \\
&= \left(240 \ \frac{\text{lbf}}{\text{ft}}\right)\left(\frac{100 \text{ ft}}{2}\right) + \left(300 \ \frac{\text{lbf}}{\text{ft}}\right)\left(\frac{60 \text{ ft}}{2}\right) \\
&= 21{,}000 \text{ lbf} \quad (21 \text{ kips})
\end{aligned}
$$

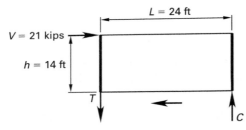

elevation of wall on line B

The overturning moment on the wall is

$$
\begin{aligned}
M_{\text{OT}} &= Vh = (21 \text{ kips})(14 \text{ ft}) \\
&= 294 \text{ ft-kips}
\end{aligned}
$$

The axial force in the shear wall boundary members is

$$
\begin{aligned}
T = C &= \frac{M_{\text{OT}}}{L} = \frac{294 \text{ ft-kips}}{24 \text{ ft}} \\
&= 12.3 \text{ kips} \quad (12 \text{ kips})
\end{aligned}
$$

The answer is (B).

124. Using the dummy load method, the unit virtual force is applied at D in the direction of the required deflection.

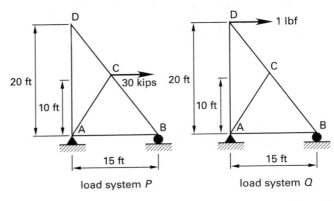

load system P   load system Q

The member forces for the real loads, load system P, and for the dummy load system, system Q, are found using basic statics.

member	N_P (kips)	N_Q (lbf)	L (in)	$N_P N_Q L$ (kips-lbf-in)
AB	15.0	1.0	180	2700
AC	25.0	0	150	0
AD	0	1.33	240	0
BC	−25.0	−1.67	150	6263
CD	0	−1.67	150	0
				8963

Applying the virtual work principle,

$$
(1 \text{ lbf})\Delta_{D,h} = \sum_{i=1}^{5} \left(\frac{N_P N_Q L}{AE}\right)_i = \frac{(8963 \text{ kips-lbf-in})}{(8 \text{ in}^2)\left(29{,}000 \ \frac{\text{kips}}{\text{in}^2}\right)}
$$

$$
\Delta_{D,h} = 0.0386 \text{ in to the right}
$$

$$
(0.04 \text{ in to the right})
$$

The answer is (C).

125. The height of the roof above the base is

$$
h_n = 14 \text{ ft} + 12 \text{ ft} = 26 \text{ ft}
$$

The building does not have moment-resulting frames, so the period can be approximated from the following formula.

$$
T = C_t h_n^{0.75} = (0.020)(26 \text{ ft})^{0.75} = 0.23 \text{ sec}
$$

The seismic dead load for NS ground motion includes the dead weight of second floor, roof, and exterior walls.

$$
\begin{aligned}
W &= (w_{D2} + w_{Dr})BL + 2w_{\text{wall}}(B+L)h \\
&= \left(30 \ \frac{\text{lbf}}{\text{ft}^2} + 20 \ \frac{\text{lbf}}{\text{ft}^2}\right)(120 \text{ ft})(60 \text{ ft}) + (2)\left(15 \ \frac{\text{lbf}}{\text{ft}^2}\right) \\
&\quad \times (60 \text{ ft} + 120 \text{ ft})\left(\frac{14 \text{ ft}}{2} + 12 \text{ ft} + 3 \text{ ft}\right) \\
&= 478{,}800 \text{ lbf} \quad (479 \text{ kips})
\end{aligned}
$$

For a building frame system consisting of light-frame walls with wood structural shear panels, $R = 7.0$. The base shear is given as

$$V = C_s W = \left(\frac{S_{DS}}{\frac{R}{I_E}}\right) W = \left(\frac{0.6}{\frac{7.0}{1.0}}\right) (479 \text{ kips})$$

$$= 41.04 \text{ kips}$$

The base shear need not be greater than

$$V = \frac{S_{D1}}{T\left(\frac{R}{I_E}\right)} = \left(\frac{0.2}{(0.23)\left(\frac{7.0}{1.0}\right)}\right) (479 \text{ kips})$$

$$= 59.4 \text{ kips}$$

In the expression above, T is the magnitude of the period and is dimensionless. The base shear must be greater than

$$V = 0.044 I_E S_{DS} W = (0.044)(1.0)(0.6)(479 \text{ kips})$$

$$= 12.6 \text{ kips}$$

Therefore, $V = 41.04 \text{ kips}$ (41 kips).

The answer is (B).

126. The influence line for the moment at D is obtained by cutting through point D and giving a small unit displacement so that only the unknown moment at D does internal work. The corresponding displaced shape is shown as the dashed line and is the influence line for the moment at D.

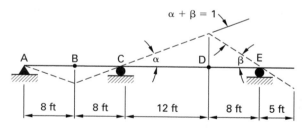

From the illustration, $\alpha + \beta = 1$. The work done by the moment through the small displacement is

$$W = M_D \alpha + M_D \beta$$

$$= M_D (\alpha + \beta)$$

$$= M_D$$

From geometry (small angle theory), the vertical displacement at point D is

$$\Delta_D = (12 \text{ ft})\alpha = (8 \text{ ft})\beta$$

$$\alpha = \frac{2\beta}{3}$$

Substituting,

$$\alpha + \beta = 1$$

$$\frac{2\beta}{3} + \beta = 1$$

$$\beta = 3/5$$

The ordinate of the influence line at point D is

$$\Delta_D = (8 \text{ ft})\beta$$

$$= (8 \text{ ft})\left(\frac{3}{5}\right)\left(1 \frac{\text{kip}}{\text{kip}}\right)$$

$$= 4.8 \text{ ft-kips/kip} \quad (5 \text{ ft-kips/kip})$$

The answer is (C).

127. The effective span length is defined in AASHTO *LRFD* Sec. 9.7.2.3 as the clear distance between flange tips plus the overhanging width of one flange.

$$S = L + \frac{b_f}{2} = (8.5 \text{ ft} - 1 \text{ ft}) + \frac{1 \text{ ft}}{2} = 8.0 \text{ ft}$$

For a concrete deck slab that is continuous over three or more supporting girders, AASHTO *LRFD* Table A4-1 gives the controlling positive and negative moments caused by a 32 kips axle load, amplified for dynamic load allowance per AASHTO *LRFD* Sec. 3.6.1.1.2, and distributed over the equivalent strip as defined in AASHTO *LRFD* Table 4.6.2.1.3-1. For positive moment in a deck with an 8.0 ft effective span, AASHTO *LRFD* Table A4-1 gives $M_{LL+IM} = 5.69$ ft-kip/ft. Thus, the factored positive moment for strength check (case I) is given by AASHTO *LRFD* Table 3.4.1-1 and Table 3.4.1-2 as

$$M_u^+ = 1.25 M_{DC} + 1.5 M_{DW} + 1.75 M_{LL+IM}$$

$$= (1.25)\left(1.0 \frac{\text{ft-kip}}{\text{ft}}\right) + (1.5)\left(0.3 \frac{\text{ft-kip}}{\text{ft}}\right)$$

$$+ (1.75)\left(5.69 \frac{\text{ft-kips}}{\text{ft}}\right)$$

$$= 11.7 \text{ ft-kips/ft} \quad (12 \text{ ft-kips/ft})$$

The answer is (C).

128. For the segment AB,

$$A_1 = \frac{\pi(D_o^2 - D_i^2)}{4} = \frac{\pi\left((3 \text{ in})^2 - (1.5 \text{ in})^2\right)}{4}$$

$$= 5.30 \text{ in}^2$$

For the segment BC,

$$A_2 = \frac{\pi D_o^2}{4} = \frac{\pi(3 \text{ in})^2}{4}$$

$$= 7.07 \text{ in}^2$$

The tension stress in segment AB is

$$f \leq F_a$$

$$\frac{P}{A} \leq F_a$$

$$\frac{P}{5.30 \text{ in}^2} \leq 22 \ \frac{\text{kips}}{\text{in}^2}$$

$$P \leq 117 \text{ kips}$$

Since the force in segment BC is smaller than in segment AB and since segment BC has a larger cross-sectional area than segment AB, BC is not the controlling section for stress. The limit on the tip deflection is

$$\Delta \leq 0.04 \text{ in}$$

$$\sum \frac{PL}{AE} \leq 0.04 \text{ in}$$

$$\frac{P(20 \text{ in})}{(5.30 \text{ in}^2)\left(29{,}000 \ \frac{\text{kips}}{\text{in}^2}\right)}$$

$$+ \frac{P(30 \text{ in})}{(7.07 \text{ in}^2)\left(29{,}000 \ \frac{\text{kips}}{\text{in}^2}\right)}$$

$$- \frac{(50 \text{ kips})(30 \text{ in})}{(7.07 \text{ in}^2)\left(29{,}000 \ \frac{\text{kips}}{\text{in}^2}\right)} \leq 0.04 \text{ in}$$

$$P \leq 171 \text{ kips}$$

The stress in the segment AB controls.

The answer is (B).

129. The influence line for shear midway between B and C, point B$'$, is obtained by cutting the beam and giving it a unit displacement such that only the shear at that point does any work. The resulting influence line is

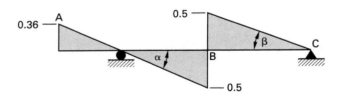

Absolute maximum shear force at B$'$ occurs with dead load over the full length and the uniform live load only over those regions where the ordinates of the influence line are positive.

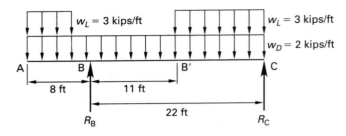

The reaction at point C is found by summing the moments about the left support. Assume clockwise moment is positive.

$$\sum M_B = 0$$

$$\left(2 \ \frac{\text{kips}}{\text{ft}}\right)(30 \text{ ft})(7 \text{ ft})$$

$$+ \left(3 \ \frac{\text{kips}}{\text{ft}}\right)\left((11 \text{ ft})(22 \text{ ft} - 5.5 \text{ ft})\right.$$

$$\left. - (8 \text{ ft})(4 \text{ ft})\right) - (22 \text{ ft})R_C = 0$$

$$R_C = 39.5 \text{ kips}$$

$$(39 \text{ kips})$$

The absolute maximum shear at B$'$ is

$$V_{B'} = \left| -R_C + w\frac{L}{2} \right| = \left| -39.5 \text{ kips} + \left(5 \ \frac{\text{kips}}{\text{ft}}\right)\left(\frac{22 \text{ ft}}{2}\right) \right|$$

$$= 15.5 \text{ kips} \quad (15 \text{ kips})$$

The answer is (B).

130. The rotation is obtained by applying a unit dummy couple at joint C and applying the virtual work principle.

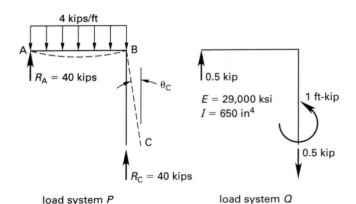

For load system P, the moment for member AB is

$$M_P = (40 \text{ kips})x - \left(4 \ \frac{\text{kips}}{\text{ft}}\right)\frac{x^2}{2}$$

$$0 \text{ ft} \leq x \leq 20 \text{ ft}$$

The moment for member BC is

$$M_P = 0$$

$$0 \text{ ft} \le x \le 20 \text{ ft}$$

To maintain equilibrium under the dummy loading, an upward force of 0.05 kip must act at A, with an equal and opposite force at C. Since the moment in member BC is zero throughout in load system P, the moment in member AB is needed only for load system Q.

$$M_Q = (0.05 \text{ kip})x$$

$$0 \text{ ft} \le x \le 20 \text{ ft}$$

The rotation at C is

$$(1 \text{ ft-kip})\theta_C = \int_L \frac{M_Q M_P \, dx}{EI}$$

$$= \int_{0 \text{ ft}}^{20 \text{ ft}} \frac{(0.05 \text{ kip})x \times \left(\begin{array}{c} (40 \text{ kips})x \\ -\left(2 \, \frac{\text{kips}}{\text{ft}}\right)x^2 \end{array} \right) dx}{EI}$$

$$\theta_C = \frac{\left(2.0 \, \frac{\text{kips}}{\text{ft}}\right)x^3}{3EI} - \frac{\left(0.1 \, \frac{\text{kip}}{\text{ft}^2}\right)x^4}{4EI} \Bigg|_{0 \text{ ft}}^{20 \text{ ft}}$$

$$= \frac{\left(\dfrac{\left(2.0 \, \frac{\text{kips}}{\text{ft}}\right)(20 \text{ ft})^3}{3} - \dfrac{\left(0.1 \, \frac{\text{kip}}{\text{ft}^2}\right)(20 \text{ ft})^4}{4} \right) \times \left(12 \, \frac{\text{in}}{\text{ft}}\right)^2}{\left(29{,}000 \, \frac{\text{kips}}{\text{in}^2}\right)(650 \text{ in}^4)}$$

$$= 0.0102 \text{ rad} \quad [\text{counterclockwise}]$$

$$(0.01 \text{ rad counterclockwise})$$

The answer is (D).

131. For $1/2$ in diameter, 270 ksi strands, the area of one strand is 0.153 in^2, and the modulus of elasticity is 28,500 ksi. The modulus of elasticity of the concrete at time of release is

$$E_c = 57{,}000\sqrt{f_c'} = 57{,}000\sqrt{3500 \, \frac{\text{lbf}}{\text{in}^2}}$$

$$= 3{,}370{,}000 \text{ lbf/in}^2 \quad (3370 \text{ kips/in}^2)$$

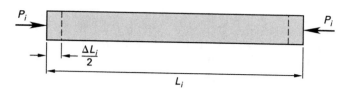

Use a trial and error method to compute loss due to elastic shortening. As a first trial, assume $\Delta f_s = 10$ ksi.

$$P_i = f_{\text{pi}} A_{\text{ps}} = \left(200 \, \frac{\text{kips}}{\text{in}^2} - 10 \, \frac{\text{kips}}{\text{in}^2}\right)(4)(0.153 \text{ in}^2)$$

$$= 116 \text{ kips}$$

Following the usual assumptions for prestressed concrete, the nominal axial stress in the concrete is based on the gross concrete area.

$$f_{\text{ci}} = \frac{P_i}{A_c} = \frac{116 \text{ kips}}{(12 \text{ in})(12 \text{ in})}$$

$$= 0.806 \text{ kip/in}^2$$

For pre-tensioned members, the change in strain in the strands is the same as in the surrounding concrete. Therefore, the computed loss of prestress is the change in strain multiplied by the modulus of elasticity.

$$\Delta f_s = \frac{f_{\text{ci}}}{E_c} E_{\text{ps}}$$

$$= \left(\frac{0.806 \, \frac{\text{kip}}{\text{in}^2}}{3370 \, \frac{\text{kips}}{\text{in}^2}} \right) \left(28{,}500 \, \frac{\text{kips}}{\text{in}^2}\right)$$

$$= 6.8 \text{ kips/in}^2$$

The actual value of Δf_s is between the trial value, 10 ksi, and the value computed using that trial value, 6.8 ksi. For a second trial, assume 7.0 ksi.

$$P_i = \left(200 \, \frac{\text{kips}}{\text{in}^2} - 7.0 \, \frac{\text{kips}}{\text{in}^2}\right)(4)(0.153 \text{ in}^2)$$

$$= 118 \text{ kips}$$

$$\Delta f_s = \frac{f_{\text{ci}}}{E_c} E_{\text{ps}}$$

$$= \left(\frac{\frac{118 \text{ kips}}{(12 \text{ in})(12 \text{ in})}}{3370 \, \frac{\text{kips}}{\text{in}^2}} \right) \left(28{,}500 \, \frac{\text{kips}}{\text{in}^2}\right)$$

$$= 6.9 \text{ kips/in}^2 \quad (7 \text{ ksi})$$

The answer is (B).

132. The pile group is subjected to combined axial compression plus biaxial bending. Maximum compression occurs in the pile farthest from the pile group centroid at the location where the forces due to bending and axial compression are additive.

$$M_x = Pe_y = (800 \text{ kips})(1.6 \text{ ft})$$
$$= 1280 \text{ ft-kips}$$
$$M_y = Pe_x = (800 \text{ kips})(1.2 \text{ ft})$$
$$= 960 \text{ ft-kips}$$

The distance from the centroid of the pile group to the first line of piles is 1.5 ft, and the distance to the second line is 4.5 ft. The centroidal moments of inertia of the pile group are

$$I_x = I_y = (2)(4)\left((4.5 \text{ ft})^2 + (1.5 \text{ ft})^2\right) = 180 \text{ ft}^2$$

$$P_{\max} = \frac{P}{n} + \frac{M_x c_x}{I_x} + \frac{M_y c_y}{I_y}$$
$$= \frac{800 \text{ kips}}{16} + \frac{(1280 \text{ ft-kips})(4.5 \text{ ft})}{180 \text{ ft}^2}$$
$$\quad + \frac{(960 \text{ ft-kips})(4.5 \text{ ft})}{180 \text{ ft}^2}$$
$$= 106 \text{ kips} \quad (110 \text{ kips})$$

The answer is (D).

133.

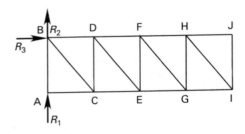

The truss has 3 external reaction components, 17 members, and 10 joints.

$$m + n_R = 2j$$
$$17 + 3 = (2)(10)$$

The necessary condition for a statically determinate truss is satisfied, but this is not sufficient to prove the truss is statically determinate. There is a problem with the support arrangement. The three reaction components are concurrent at joint B. If a nonconcurrent force is applied at any joint, it is impossible to satisfy the equilibrium condition that the summation of moments about joint B must equal zero. Therefore, the truss is unstable.

The answer is (B).

134. For a thin-walled pipe with internal pressure and negligible longitudinal restraint, the maximum stress occurs in the circumferential direction and can be calculated using the membrane theory of thin-walled pressure vessels.

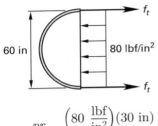

$$f_t = \frac{pr}{t} = \frac{\left(80 \, \frac{\text{lbf}}{\text{in}^2}\right)(30 \text{ in})}{0.375 \text{ in}}$$
$$= 6400 \text{ lbf/in}^2$$

Since the longitudinal direction is unrestrained, the strain in the circumferential direction is found from the uniaxial case.

$$\epsilon = \frac{f_t}{E} = \frac{6400 \, \frac{\text{lbf}}{\text{in}^2}}{29{,}000{,}000 \, \frac{\text{lbf}}{\text{in}^2}} = 0.00022$$

The change in diameter is directly proportional to the change in circumference.

$$\Delta D = \frac{\epsilon \pi D}{\pi} = (0.00022)(60 \text{ in})$$
$$= 0.0132 \text{ in} \quad (0.013 \text{ in})$$

The answer is (B).

135. Release the reaction at point B to create a stable determinate beam.

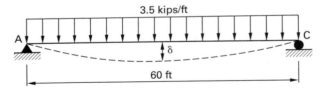

The downward displacement in the released structure at B caused by the applied uniformly distributed load is

$$\delta = \frac{5wL^4}{384EI} = \frac{(5)\left(3.5 \, \frac{\text{kips}}{\text{ft}}\right)(60 \text{ ft})^4\left(12 \, \frac{\text{in}}{\text{ft}}\right)^3}{(384)\left(29{,}000 \, \frac{\text{kips}}{\text{in}^2}\right)(1630 \text{ in}^4)}$$
$$= 21.59 \text{ in}$$

The flexibility coefficient, f, is obtained by applying a unit force upward at the released point and computing the displacement caused by that force at the released point.

$$f = \frac{L^3}{48EI} = \frac{\left((60 \text{ ft})\left(12 \frac{\text{in}}{\text{ft}}\right)\right)^3}{(48)\left(29{,}000 \frac{\text{kips}}{\text{in}^2}\right)(1630 \text{ in}^4)}$$

$$= 0.1645 \text{ in/kip}$$

For consistent displacement at point B,

$$\delta + R_{\text{B}}f = -0.5 \text{ in}$$

$$-21.59 \text{ in} + R_{\text{B}}\left(0.1645 \frac{\text{in}}{\text{kip}}\right) = -0.5 \text{ in}$$

$$R_{\text{B}} = 128 \text{ kips} \quad (130 \text{ kips})$$

The answer is (C).

136. Due to symmetry, there are only two possible collapse mechanisms. In the first mechanism, plastic hinges form in the end spans.

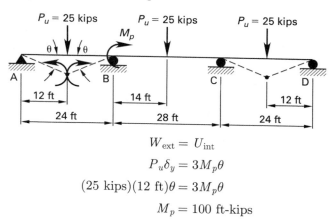

$$W_{\text{ext}} = U_{\text{int}}$$

$$P_u\delta_y = 3M_p\theta$$

$$(25 \text{ kips})(12 \text{ ft})\theta = 3M_p\theta$$

$$M_p = 100 \text{ ft-kips}$$

In the second mechanism, plastic hinges form in the interior span.

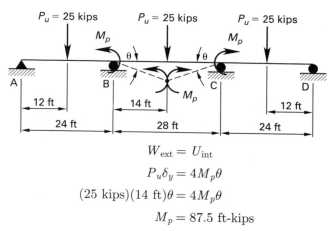

$$W_{\text{ext}} = U_{\text{int}}$$

$$P_u\delta_y = 4M_p\theta$$

$$(25 \text{ kips})(14 \text{ ft})\theta = 4M_p\theta$$

$$M_p = 87.5 \text{ ft-kips}$$

Collapse is controlled by the exterior spans, so $M_p = 100$ ft-kips.

The answer is (D).

137. Let K represent the force applied at midspan producing a unit midspan deflection.

$$\frac{KL^3}{48EI} = 1$$

$$K = \frac{48EI}{L^3}$$

$$= \frac{(48)\left(29{,}000 \frac{\text{kips}}{\text{in}^2}\right)(1600 \text{ in}^4)}{\left((16 \text{ ft})\left(12 \frac{\text{in}}{\text{ft}}\right)\right)^3}$$

$$= 315 \text{ kips/in}$$

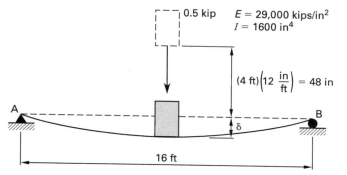

Conservation of energy requires that the potential energy of the 0.5 kip weight is converted into strain energy in the beam.

$$E_p = Wy = (0.5 \text{ kip})(48 \text{ in} + \delta)$$

$$U_{\text{int}} = 0.5(K\delta)\delta = (0.5)\left(315 \frac{\text{kips}}{\text{in}}\right)\delta^2$$

Equating the two energy terms gives

$$E_p = U_{\text{int}}$$

$$(0.5 \text{ kip})(48 \text{ in} + \delta) = (0.5)\left(315 \frac{\text{kips}}{\text{in}}\right)\delta^2$$

$$\left(315 \frac{\text{kips}}{\text{in}}\right)\delta^2 - \delta - 48 \text{ in} = 0$$

$$\delta = 0.392 \text{ in}$$

The falling weight causes a maximum midspan deflection of 0.392 in, which corresponds to a peak force of

$$F = K\delta = \left(315 \frac{\text{kips}}{\text{in}}\right)(0.392 \text{ in})$$

$$= 123.5 \text{ kips} \quad (120 \text{ kips})$$

The answer is (D).

138. Let K represent the force applied at the top of each column producing a unit horizontal deflection.

$$\frac{KH^3}{12EI} = 1$$

$$K = \frac{12EI}{H^3} = \frac{(12)\left(29{,}000\ \dfrac{\text{kips}}{\text{in}^2}\right)(800\ \text{in}^4)}{\left((16\ \text{ft})\left(12\ \dfrac{\text{in}}{\text{ft}}\right)\right)^3}$$

$$= 39.3\ \text{kips/in} \quad \text{[each column]}$$

The equivalent stiffness of the four supporting columns is four times that of an individual column. The natural period of vibration of the system is

$$T = 2\pi\sqrt{\frac{m}{4K}} = 2\pi\sqrt{\frac{\dfrac{W}{g}}{4K}}$$

$$= 2\pi\sqrt{\frac{30\ \text{kips}}{\left(32.2\ \dfrac{\text{ft}}{\text{sec}^2}\right)(4)\left(39.3\ \dfrac{\text{kips}}{\text{in}}\right)\left(12\ \dfrac{\text{in}}{\text{ft}}\right)}}$$

$$= 0.14\ \text{sec} \quad (0.15\ \text{sec})$$

The answer is (A).

139. Relevant properties of the W12 × 106 include its area, 31.2 in²; the moment of inertia about its weak axis, 301 in⁴; the moment of inertia about its strong axis, 933 in⁴; and its overall depth, 12.9 in.

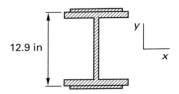

The centroid of the section is on the axes of symmetry and is located by inspection. Properties of the built-up section are

$$A = \sum A_i = 31.2\ \text{in}^2 + (2)(10\ \text{in})(0.625\ \text{in}) = 43.7\ \text{in}^2$$

$$I_x = \sum(I_{xc} + Ad^2)$$

$$= 933\ \text{in}^4 + (2)\left(\begin{array}{c}\dfrac{(10\ \text{in})(0.625\ \text{in})^3}{12} \\ + (10\ \text{in})(0.625\ \text{in}) \\ \times \left(\dfrac{12.9\ \text{in} + 0.625\ \text{in}}{2}\right)^2\end{array}\right)$$

$$= 1505\ \text{in}^4$$

$$I_y = \sum I_{yc} = 301\ \text{in}^4 + (2)\left(\dfrac{(0.625\ \text{in})(10\ \text{in})^3}{12}\right)$$

$$= 405\ \text{in}^4$$

The major principal axis is the x-axis. The radius of gyration is

$$r_x = \sqrt{\frac{I_x}{A}} = \sqrt{\frac{1505\ \text{in}^4}{43.7\ \text{in}^2}}$$

$$= 5.87\ \text{in} \quad (5.9\ \text{in})$$

The answer is (A).

140. The relevant properties of a W24 × 55 are its area, 16.3 in²; the moment of inertia about its x-axis, 1360 in⁴; and its overall depth, 23.6 in. The centroid of the built-up section is located by inspection from symmetry.

$$I_x = \sum(I_{xc} + Ad^2)$$

$$= 1360\ \text{in}^4 + (2)\left(\begin{array}{c}\dfrac{(8\ \text{in})(0.5\ \text{in})^3}{12} + (8\ \text{in})(0.5\ \text{in}) \\ \times \left(\dfrac{23.6\ \text{in} + 0.5\ \text{in}}{2}\right)^2\end{array}\right)$$

$$= 2522\ \text{in}^4$$

The shear flow between the coverplate and flange depends on the statical moment, Q, of the coverplate area about the neutral axis of bending.

$$Q = A\bar{y}$$

$$= (8\ \text{in})(0.5\ \text{in})\left(\dfrac{23.6\ \text{in} + 0.5\ \text{in}}{2}\right)$$

$$= 48.2\ \text{in}^3$$

$$q = \frac{VQ}{I}$$

$$= \frac{(95\ \text{kips})(48.2\ \text{in}^3)}{2522\ \text{in}^4}$$

$$= 1.8\ \text{kips/in} \quad (2\ \text{kips/in})$$

The answer is (A).

141.

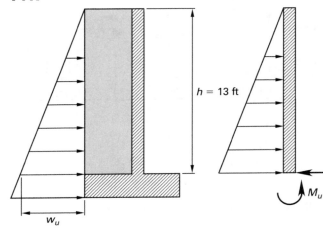

Design the stem on a per-foot-of-width basis using ACI 318. The design lateral load at the base of the stem is

$$w_u = 1.6\gamma_a h(1 \text{ ft})$$

$$= (1.6)\left(45 \frac{\text{lbf}}{\text{ft}^3}\right)(13 \text{ ft})(1 \text{ ft})$$

$$= 936 \text{ lbf/ft}$$

The design bending moment at the base is

$$M_u = \frac{0.5 w_u h^2}{3}$$

$$= \frac{(0.5)\left(936 \frac{\text{lbf}}{\text{ft}}\right)(13 \text{ ft})^2}{3}$$

$$= 26{,}364 \text{ ft-lbf}$$

For a unit width of wall ($b = 12$ in) and a specified steel percentage of 0.01,

$$M_u = \phi M_n = \phi \rho f_y b d^2\left(1 - 0.59\rho \frac{f_y}{f'_c}\right)$$

Inserting known values and equating to the value of M_u gives the required wall thickness, d.

$$(26{,}364 \text{ ft-lbf})$$

$$\times\left(12 \frac{\text{in}}{\text{ft}}\right) = (0.9)(0.01)\left(60{,}000 \frac{\text{lbf}}{\text{in}^2}\right)(12 \text{ in})d^2$$

$$\times\left(1 - (0.59)(0.01)\left(\frac{60{,}000 \frac{\text{lbf}}{\text{in}^2}}{4000 \frac{\text{lbf}}{\text{in}^2}}\right)\right)$$

$$d = 7.32 \text{ in}$$

The steel area is

$$A_s = \rho b d = (0.01)\left(12 \frac{\text{in}}{\text{ft}}\right)(7.32 \text{ in})$$

$$= 0.88 \text{ in}^2/\text{ft} \quad (0.90 \text{ in}^2/\text{ft})$$

The answer is (D).

142. According to ACI 318 Table 4.3.1, reinforced normalweight concrete exposed to seawater (exposure class C2) is limited to a maximum water-to-cementitious-materials ratio of 0.40.

The answer is (B).

143. AASHTO *LRFD* Sec. 5.4.2.3.2 describes the method of determining creep.

Calculate the volume, V, of the girder.

$$V = LDW$$

$$= (45 \text{ ft})(3.0 \text{ ft})(1.5 \text{ ft})$$

$$= 202.5 \text{ ft}^3$$

Calculate the total surface area, S, of the girder.

$$S = 2(LD + DW + LW)$$

$$= (2)\left(\begin{array}{c}(45 \text{ ft})(3.0 \text{ ft}) + (3.0 \text{ ft})(1.5 \text{ ft}) \\ + (45 \text{ ft})(1.5 \text{ ft})\end{array}\right)$$

$$= 414 \text{ ft}^2$$

Calculate the volume-to-surface area ratio.

$$\frac{V}{S} = \left(\frac{205 \text{ ft}^3}{414 \text{ ft}^2}\right)\left(12 \frac{\text{in}}{\text{ft}}\right)$$

$$= 5.94 \text{ in} \quad [\text{Use 6 in.}]$$

Use AASHTO *LRFD* Eq. 5.4.2.3.2-2 to find the factor for the effect of the volume-to-surface area ratio, k_s. k_s must be at least 1.0.

$$k_s = 1.45 - 0.13\left(\frac{V}{S}\right)$$

$$= 1.45 - (0.13)(6 \text{ in})$$

$$= 0.67 \quad [< 1.0. \text{ Use 1.0 in.}]$$

Use AASHTO *LRFD* Eq. 5.4.2.3.2-3 to find the humidity factor for creep, k_{hc}.

$$k_{hc} = 1.56 - 0.008H$$

$$= 1.56 - (0.008)(80\%)$$

$$= 0.92$$

Use AASHTO *LRFD* Eq. 5.4.2.3.2-4 to find the factor for the effect of concrete strength, k_f.

$$k_f = \frac{5}{1 + f'_{c,i}} = \frac{5}{1 + 4 \frac{\text{kips}}{\text{in}^2}}$$

$$= 1.0$$

Use AASHTO *LRFD* Eq. 5.4.2.3.2-5 to find the time-development factor, k_{td}.

$$k_{td} = \frac{t}{61 - 4f'_{c,i} + t}$$

$$= \frac{45 \text{ days} - 35 \text{ days}}{61 - (4)\left(4 \frac{\text{kips}}{\text{in}^2}\right) + (45 \text{ days} - 35 \text{ days})}$$

$$= 0.182$$

From AASHTO *LRFD* Eq. 5.4.2.3.2-1, the creep coefficient, Ψ, is

$$\Psi = 1.9 k_s k_{hc} k_f k_{td} t_i^{-0.118}$$

$$= (1.9)(1.0)(0.92)(1.0)(0.182)(45 \text{ days})^{-0.118}$$

$$= 0.2$$

The answer is (B).

144. Balanced strain conditions exist when the strain in the steel on the tension side reaches yield and, at the same time, the strain in the extreme compression edge of the concrete is at the ultimate value of 0.003.

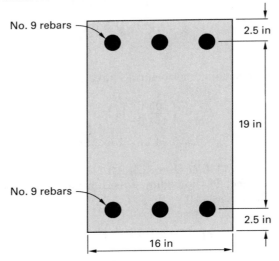

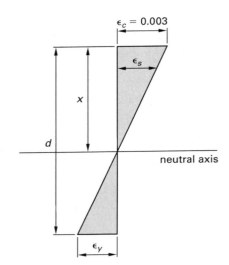

The yield strain for grade 60 rebar is

$$\epsilon_y = \frac{f_y}{E_s} = \frac{60 \frac{\text{kips}}{\text{in}^2}}{29{,}000 \frac{\text{kips}}{\text{in}^2}} = 0.00207$$

From similar triangles,

$$\frac{x}{0.003} = \frac{d}{0.003 + \epsilon_y}$$

$$x = \frac{(21.5 \text{ in})(0.003)}{0.003 + 0.00207} = 12.7 \text{ in}$$

$$\frac{\epsilon'_s}{x - 2.5 \text{ in}} = \frac{0.003}{x}$$

$$\epsilon'_s = \frac{(12.7 \text{ in} - 2.5 \text{ in})(0.003)}{12.7 \text{ in}} = 0.0024 > \epsilon_y$$

The strain in the compression steel exceeds yield. Thus, the stress in the compression steel is the yield stress, and the force in the compression steel is

$$C'_s = f_y A'_s = \left(60 \frac{\text{kips}}{\text{in}^2} \right)(3.00 \text{ in}^2) = 180 \text{ kips}$$

The answer is (D).

145. For a concentrically loaded tied column, the design strength is given by

$$\phi P_{n,\max} = 0.8\phi(0.85f'_c(A_g - A_{\text{st}}) + f_y A_{\text{st}})$$

For a specified longitudinal steel ratio of 0.02,

$$A_{\text{st}} = \rho_g A_g = 0.02 A_g$$

Substituting gives

$$1090 \text{ kips} = \phi P_{n,\max}$$
$$= (0.8)(0.65)$$
$$\times \left(\begin{array}{c} (0.85)\left(5 \frac{\text{kips}}{\text{in}^2} \right)(A_g - 0.02 A_g) \\ + \left(60 \frac{\text{kips}}{\text{in}^2} \right)(0.02 A_g) \end{array} \right)$$
$$A_g = 391 \text{ in}^2$$
$$b = \sqrt{A_g} = \sqrt{391 \text{ in}^2} = 19.8 \text{ in} \quad (20 \text{ in})$$

The answer is (C).

146. Use the appropriate reinforced concrete interaction diagram from *Design of Concrete Structures*, or the equivalent. For a column with a 20 in cross section in the direction that resists bending, the steel placement constant, γ, is found using

$$\gamma h = h - 2d'$$
$$\gamma(20 \text{ in}) = 20 \text{ in} - (2)(3 \text{ in}) = 14 \text{ in}$$
$$\gamma = 0.7$$

The reference interaction diagram requires two parameters.

$$R_n = \frac{P_u e}{\phi f'_c A_g h} = \frac{M_u}{\phi f'_c A_g h}$$
$$= \frac{(175 \text{ ft-kips})\left(12 \frac{\text{in}}{\text{ft}} \right)}{(0.65)\left(4 \frac{\text{kips}}{\text{in}^2} \right)(20 \text{ in})(18 \text{ in})(20 \text{ in})}$$
$$= 0.11$$

$$K_n = \frac{P_u}{\phi f'_c A_g} = \frac{875 \text{ kips}}{(0.65)\left(4 \frac{\text{kips}}{\text{in}^2} \right)(20 \text{ in})(18 \text{ in})} = 0.93$$

From the interaction curves, interpolation at the point (0.11, 0.93) gives a longitudinal steel ratio, ρ_g, of 0.022. The required steel area is

$$A_{\text{st}} = \rho_g A_g = (0.022)(20 \text{ in})(18 \text{ in}) = 7.92 \text{ in}^2 \quad (8 \text{ in}^2)$$

The answer is (B).

147. The development length can be calculated from the equations given in either ACI 318 Sec. 12.2.2 or Sec. 12.2.3. The simplified equations in ACI 318 Sec. 12.2.2 incorporate default values in the calculation of the confinement term, $(c_b + K_{\text{tr}})/d_b$, which result in more conservative (i.e., larger) development lengths than those calculated using ACI 318 Eq. 12-1 and Eq. 12-2. To determine the minimum development length, use the equations in ACI 318 Sec. 12.2.3.

There are four possible initiating cracks, which are labeled as c_{b1}, c_{b2}, c_{b3}, and c_{b4}. The initiating crack represents the location where the beam will most likely split first and is indicated by the smallest c_b value. c_{b1} is the horizontal distance from the center of the bar to the nearest concrete surface; c_{b2} is one-half of the horizontal center-to-center spacing; c_{b3} is one-half of the vertical center-to-center spacing; and c_{b4} is the vertical distance from the tension bar to the nearest horizontal surface. Cracks c_{b1} and c_{b2} are associated with horizontal splitting planes, and cracks c_{b3} and c_{b4} are associated with vertical splitting planes.

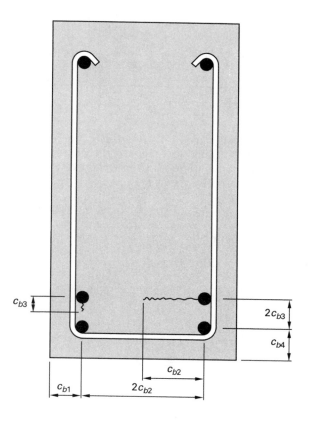

The initiating crack is

$$c_b = \min \begin{cases} c_{b1} = \text{cover} + d_{b,\text{no.3}} + \dfrac{d_{b,\text{no.9}}}{2} \\[4pt] \quad = 1.5 \text{ in} + 0.375 \text{ in} + \dfrac{1.128 \text{ in}}{2} \\[4pt] \quad = 2.439 \text{ in} \\[4pt] c_{b2} = \dfrac{9.122 \text{ in}}{2} = 4.561 \text{ in} \\[4pt] c_{b3} = \dfrac{2.25 \text{ in}}{2} = 1.125 \text{ in} \quad \text{[controls]} \\[4pt] c_{b4} = 26 \text{ in} - 21.25 \text{ in} - 2.25 \text{ in} = 2.5 \text{ in} \end{cases}$$

Since c_{b3} controls, the initiating crack will most likely occur along the vertical center-to-center distance between the bars, resulting in a vertical splitting plane as shown. While c_{b3} may occur in four different locations and result in two vertical splitting planes (shown as planes A and B), the dimensions of the planes are the same.

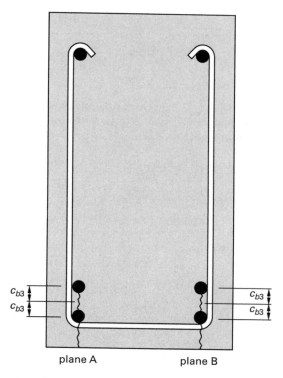

plane A   plane B

There is only one stirrup leg along the probable splitting plane. Therefore, the total transverse reinforcement area within a spacing of s is equivalent to the area of one no. 3 stirrup, 0.11 in^2. There are two bars that develop along the vertical splitting plane.

Although ACI 318 Sec. 12.2.3 allows the use of $K_{\text{tr}} = 0$, using zero will result in a development length larger than the minimum. From ACI 318 Eq. 12-2, the transverse reinforcement index is

$$K_{\text{tr}} = \frac{40A_{\text{tr}}}{sN} = \frac{(40)(0.11 \text{ in}^2)}{(3 \text{ in})(2)} = 0.733 \text{ in}$$

From ACI 318 Sec. 12.2.3, the confinement term is

$$\frac{c_b + K_{tr}}{d_b} = \frac{1.125 \text{ in} + 0.733 \text{ in}}{1.128 \text{ in}}$$
$$= 1.65 \quad [\leq 2.5, \text{ OK}]$$

Determine the adjustment factor values using ACI 318 Sec. 12.2.4. Since only 1.5 in (< 12 in) of concrete is cast below the lowest row of no. 9 bars (i.e., the bars are bottom bars, not top bars), the reinforcement location factor, ψ_t, is 1.0. For uncoated bars, the epoxy coating factor, ψ_e, is 1.0. The reinforcement size factor, ψ_s, is 1.0 for no. 7 and larger bars. The lightweight aggregate factor, λ, is 1.0 for normalweight concrete.

Using ACI 318 Eq. 12-1, calculate the development length.

$$l_d = \frac{3 d_b f_y \psi_t \psi_e \psi_s}{40 \lambda \sqrt{f'_c} \left(\dfrac{c_b + K_{tr}}{d_b} \right)}$$

$$= \frac{(3)(1.128 \text{ in}) \left(60{,}000 \ \dfrac{\text{lbf}}{\text{in}^2} \right)(1.0)(1.0)(1.0)}{(40)(1.0)\sqrt{3000 \ \dfrac{\text{lbf}}{\text{in}^2}}(1.65)}$$

$$= 56.25 \text{ in} \quad [\geq 12 \text{ in}, \text{ OK}]$$

From ACI 318 Sec. 12.2.5, the development length can be reduced by the ratio $A_{s,req}/A_{s,prov}$. Therefore, the minimum development length is

$$l_{d,\min} = \left(\frac{A_{s,req}}{A_{s,prov}} \right) l_d$$

$$= (0.74)(56.25 \text{ in})$$

$$= 41.6 \text{ in} \quad (42 \text{ in}) \quad [\geq 12 \text{ in}, \text{ OK}]$$

The answer is (B).

148. The equivalent force system acting at the centroid of the footing is

$$P = P_1 + P_2 + wbht$$
$$= 400 \text{ kips} + 200 \text{ kips}$$
$$\quad + \left(0.15 \ \frac{\text{kip}}{\text{ft}^3} \right)(26 \text{ ft})(6 \text{ ft})(2 \text{ ft})$$
$$= 647 \text{ kips}$$

$$e = \frac{P_1 e_1 + P_2 e_2}{P}$$
$$= \frac{(400 \text{ kips})(-10 \text{ ft}) + (200 \text{ kips})(12.5 \text{ ft})}{647 \text{ kips}}$$
$$= -2.3 \text{ ft}$$

Since the resultant force on the footing acts within the middle third of the footing's length, the entire area beneath the footing is in compression.

$$A = bh = (6 \text{ ft})(26 \text{ ft}) = 156 \text{ ft}^2$$

$$S = \frac{hb^2}{6} = \frac{(6 \text{ ft})(26 \text{ ft})^2}{6}$$
$$= 676 \text{ ft}^3$$

$$f_{p,\max} = \frac{P}{A} + \frac{Pe}{S}$$
$$= \frac{647 \text{ kips}}{156 \text{ ft}^2} + \frac{(647 \text{ kips})(2.3 \text{ ft})}{676 \text{ ft}^3}$$
$$= 6.3 \text{ kips/ft}^2 \quad (6 \text{ kips/ft}^2)$$

The answer is (D).

149.

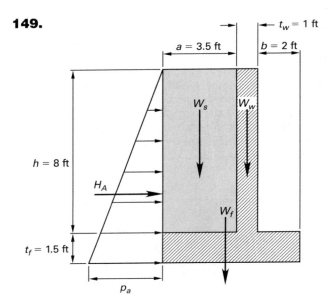

Per unit length of wall (1 ft), the force components resisting overturning are the weight of the soil, the weight of the wall stem, and the weight of the footing. The force exerted by the soil is

$$W_s = \gamma_s a h(1 \text{ ft}) = \left(100 \ \frac{\text{lbf}}{\text{ft}^3} \right)(3.5 \text{ ft})(8 \text{ ft})(1 \text{ ft})$$
$$= 2800 \text{ lbf}$$

The force exerted by the wall stem is

$$W_w = \gamma_c t_w h(1 \text{ ft}) = \left(150 \ \frac{\text{lbf}}{\text{ft}^3} \right)(1 \text{ ft})(8 \text{ ft})(1 \text{ ft})$$
$$= 1200 \text{ lbf}$$

The force exerted by the footing is

$$W_f = \gamma_c t_f (a + t_w + b)(1 \text{ ft})$$
$$= \left(150 \ \frac{\text{lbf}}{\text{ft}^3} \right)(1.5 \text{ ft})(3.5 \text{ ft} + 1 \text{ ft} + 2 \text{ ft})(1 \text{ ft})$$
$$= 1463 \text{ lbf}$$

The resisting moment about the toe of the footing is

$$M_r = \sum W_i x_i$$
$$= (2800 \text{ lbf})(4.75 \text{ ft}) + (1200 \text{ lbf})(2.5 \text{ ft})$$
$$+ (1463 \text{ lbf})(3.25 \text{ ft})$$
$$= 21{,}055 \text{ ft-lbf}$$

The active earth pressure at the base is

$$p_a = \gamma_f(h + t_f)(1 \text{ ft}) = \left(35 \ \frac{\text{lbf}}{\text{ft}^3}\right)(8 \text{ ft} + 1.5 \text{ ft})(1 \text{ ft})$$
$$= 333 \text{ lbf/ft}$$

The total active earth pressure is

$$H_A = 0.5 p_a(h + t_f) = (0.5)\left(333 \ \frac{\text{lbf}}{\text{ft}}\right)(8 \text{ ft} + 1.5 \text{ ft})$$
$$= 1582 \text{ lbf}$$

The overturning moment is

$$M_o = H_A\left(\frac{h + t_f}{3}\right) = (1582 \text{ lbf})\left(\frac{8 \text{ ft} + 1.5 \text{ ft}}{3}\right)$$
$$= 5010 \text{ ft-lbf}$$

The factor of safety is

$$\text{FS} = \frac{M_r}{M_o} = \frac{21{,}055 \text{ ft-lbf}}{5010 \text{ ft-lbf}}$$
$$= 4.2 \quad (4)$$

The answer is (D).

150. Use the customary nomenclature and methods of analysis of the *PCI Design Handbook*. Replace the prestress by the statically equivalent force system acting at the centroid of concrete. The eccentricity, e, of the strands is their distance from midheight, or $(34 \text{ in}/2) - 2 \text{ in} = 15 \text{ in}$.

$$P_i = f_{pi} A_{ps} = \left(180 \ \frac{\text{kips}}{\text{in}^2}\right)(0.918 \text{ in}^2)$$
$$= 165 \text{ kips}$$
$$M_i = P_i e = (165 \text{ kips})(15 \text{ in})$$
$$= 2475 \text{ in-kips}$$
$$E_{ci} = 33 w_c^{1.5} \sqrt{f'_{ci}}$$
$$= (33)\left(110 \ \frac{\text{lbf}}{\text{ft}^3}\right)^{1.5} \sqrt{3500 \ \frac{\text{lbf}}{\text{in}^2}}$$
$$= 2{,}250{,}000 \text{ lbf/in}^2$$
$$I = \frac{bh^3}{12} = \frac{(14 \text{ in})(34 \text{ in})^3}{12}$$
$$= 45{,}854 \text{ in}^4$$

$$A = bh = \frac{(14 \text{ in})(34 \text{ in})}{\left(12 \ \frac{\text{in}}{\text{ft}}\right)^2}$$
$$= 3.31 \text{ ft}^2$$
$$w = w_c A = \left(110 \ \frac{\text{lbf}}{\text{ft}^3}\right)(3.31 \text{ ft}^2)$$
$$= 364 \text{ lbf/ft}$$

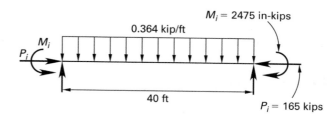

The midspan camber is the algebraic sum of the deflections caused by the end moment and transverse beam weight.

$$\delta_i = \frac{5wL^4}{384 E_{ci} I} - \frac{M_i L^2}{8 E_{ci} I}$$

$$= \frac{(5)\left(0.364 \ \frac{\text{kip}}{\text{ft}}\right)(40 \text{ ft})^4 \left(12 \ \frac{\text{in}}{\text{ft}}\right)^3}{(384)\left(2250 \ \frac{\text{kips}}{\text{in}^2}\right)(45{,}854 \text{ in}^4)}$$

$$- \frac{(2475 \text{ in-kips})(40 \text{ ft})^2 \left(12 \ \frac{\text{in}}{\text{ft}}\right)^2}{(8)\left(2250 \ \frac{\text{kips}}{\text{in}^2}\right)(45{,}854 \text{ in}^4)}$$

$$= 0.203 \text{ in} - 0.691 \text{ in}$$
$$= -0.488 \text{ in} \quad (0.5 \text{ in} \uparrow)$$

The answer is (B).

151. From AASHTO *LRFD* Sec. 3.6.3, the centrifugal force calculation requires the front axle weight of the design truck, $w_{\text{front axle}}$, and the factor C. From AASHTO *LRFD* Fig. 3.6.1.2.2-1, the front axle weight, $w_{\text{front axle}}$, of the design truck is 8 kips.

From AASHTO *LRFD* Sec. 3.6.3, $f = \frac{4}{3}$ for all load combinations other than fatigue.

Use AASHTO *LRFD* Eq. 3.6.3-1 to calculate the factor C for a radius of curvature, R, of 400 ft.

$$C = \frac{f v^2}{gR} = \frac{\left(\frac{4}{3}\right)\left(\dfrac{\left(45 \ \frac{\text{mi}}{\text{hr}}\right)\left(5280 \ \frac{\text{ft}}{\text{mi}}\right)}{\left(60 \ \frac{\text{min}}{\text{hr}}\right)\left(60 \ \frac{\text{sec}}{\text{min}}\right)}\right)^2}{\left(32.2 \ \frac{\text{ft}}{\text{sec}^2}\right)(400 \text{ ft})}$$

$$= 0.45$$

From AASHTO *LRFD* Table 3.6.1.1.2-1, for one loaded lane, the multiple-lane presence factor, m, is 1.20. From AASHTO *LRFD* Table 3.4.1-1, for a strength I load combination, the load factor, CE, is 1.75.

Calculate the factored centrifugal force, F.

$$F = C w_{\text{front axle}} m(\text{CE})$$
$$= (0.45)(8 \text{ kips})(1.20)(1.75)$$
$$= 7.56 \text{ kips} \quad (7.6 \text{ kips})$$

The answer is (D).

152.

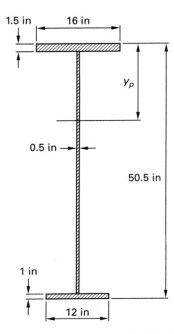

The section consists of three rectangles having areas A_1, A_2, and A_3.

$$A_i = b_i h_i$$
$$A_1 = (16 \text{ in})(1.5 \text{ in})$$
$$= 24 \text{ in}^2$$
$$A_2 = (0.5 \text{ in})(48 \text{ in})$$
$$= 24 \text{ in}^2$$
$$A_3 = (12 \text{ in})(1 \text{ in})$$
$$= 12 \text{ in}^2$$

For the fully plastic condition, the plastic neutral axis is positioned such that the area in compression equals the area in tension.

$$A_{\text{comp}} = A_{\text{ten}} = \frac{\sum A_i}{2}$$
$$= \frac{24 \text{ in}^2 + 24 \text{ in}^2 + 12 \text{ in}^2}{2}$$
$$= 30 \text{ in}^2$$

$$y_p = 1.5 \text{ in} + \frac{A_{\text{comp}} - A_1}{b_2} = 1.5 \text{ in} + \frac{30 \text{ in}^2 - 24 \text{ in}^2}{0.5 \text{ in}}$$
$$= 13.5 \text{ in}$$

The region above the plastic neutral axis is under uniform compression stress, F_y, and the region below is under uniform tension stress, which is also F_y. Taking moments of the stress resultants gives the plastic moment capacity. For convenience, take the moments about the tension force in the bottom flange.

$$M_p = \sum F_y A_i \overline{y}_i$$
$$= \frac{\left(36 \dfrac{\text{kips}}{\text{in}^2}\right) \left(\begin{array}{c} (16 \text{ in})(1.5 \text{ in})(50 \text{ in} - 0.75 \text{ in}) \\ + (0.5 \text{ in})(12 \text{ in})(50 \text{ in} - 7.5 \text{ in}) \\ - (0.5 \text{ in})(36 \text{ in})(18.5 \text{ in}) \end{array} \right)}{12 \dfrac{\text{in}}{\text{ft}}}$$
$$= 3312 \text{ ft-kips} \quad (3300 \text{ ft-kips})$$

The answer is (C).

153. The design load is controlled by

$$w_u = 1.2 w_D + 1.6 w_L$$
$$= (1.2)\left(2.5 \dfrac{\text{kips}}{\text{ft}}\right) + (1.6)\left(1.8 \dfrac{\text{kips}}{\text{ft}}\right)$$
$$= 5.88 \text{ kips/ft}$$

For a uniformly loaded simple beam,

$$M_{u,x} = \frac{w_u L x}{2} - \frac{w_u x^2}{2}$$

$$M_{\text{max}} = \frac{w_u L^2}{8} = \frac{\left(5.88 \dfrac{\text{kips}}{\text{ft}}\right)(36 \text{ ft})^2}{8}$$
$$= 953 \text{ ft-kips}$$

$$M_A = \frac{\left(5.88 \dfrac{\text{kips}}{\text{ft}}\right)(36 \text{ ft})(4.5 \text{ ft})}{2}$$
$$- \frac{\left(5.88 \dfrac{\text{kips}}{\text{ft}}\right)(4.5 \text{ ft})^2}{2}$$
$$= 417 \text{ ft-kips}$$

$$M_B = \frac{\left(5.88 \dfrac{\text{kips}}{\text{ft}}\right)(36 \text{ ft})(9 \text{ ft})}{2}$$
$$- \frac{\left(5.88 \dfrac{\text{kips}}{\text{ft}}\right)(9 \text{ ft})^2}{2}$$
$$= 714 \text{ ft-kips}$$

$$M_C = \frac{\left(5.88 \ \frac{\text{kips}}{\text{ft}}\right)(36 \ \text{ft})(13.5 \ \text{ft})}{2}$$
$$- \frac{\left(5.88 \ \frac{\text{kips}}{\text{ft}}\right)(13.5 \ \text{ft})^2}{2}$$
$$= 893 \ \text{ft-kips}$$

The lateral-torsional buckling modification factor [*AISC Specification* Eq. F1-1] is

$$C_b = \frac{12.5 M_{\max}}{2.5 M_{\max} + 3 M_A + 4 M_B + 3 M_C}$$
$$= \frac{(12.5)(953 \ \text{ft-kips})}{(2.5)(953 \ \text{ft-kips}) + (3)(417 \ \text{ft-kips})}$$
$$+ (4)(714 \ \text{ft-kips}) + (3)(893 \ \text{ft-kips})$$
$$= 1.30 \quad (1.3)$$

The answer is (D).

154. The strength of the backing plate is $F_{\text{BM}}A_{\text{BM}}$ and the strength of the weld metal is $F_W A_W$. To find the minimum thickness, set these two strengths equal to each other. The electrode is E70, so its strength is 70 ksi.

"Full strength" is interpreted as "strength to rupture." From AISC *Specification* Sec. J4.2, the factor of safety for shear rupture of the weld material, Ω_W is 2.00. However, the connection will fail first in yield of the base material. From AISC *Specification* Sec. J4.2, the factor of safety for shear yield of the base material, Ω_{BM} is 1.50.

$$\Omega_{\text{BM}} F_{\text{BM}} A_{\text{BM}} = \Omega_W F_W A_W$$
$$(1.50)(0.6 F_y)(t L_w) = (2.0)(0.6 F_{\text{EXX}})(0.707 w L_w)$$
$$t = \frac{\Omega_W F_{\text{EXX}}(0.707 w)}{\Omega_W F_y}$$
$$= \frac{(2.00)\left(70 \ \frac{\text{kips}}{\text{in}^2}\right)(0.707)}{\times (0.25 \ \text{in})(2 \ \text{welds})}{(1.5)\left(36 \ \frac{\text{kips}}{\text{in}^2}\right)}$$
$$= 0.92 \ \text{in} \quad (0.9 \ \text{in})$$

The answer is (D).

155. The net area is the gross area less the area of four flange holes and two web holes. Since holes are punched, the hole diameter is taken as $1/8$ in greater than the fastener diameter.

$$A_n = A_g - 4 t_f D - 2 t_w D$$
$$= 14.4 \ \text{in}^2 - (4)(0.56 \ \text{in})(0.875 \ \text{in})$$
$$- (2)(0.34 \ \text{in})(0.875 \ \text{in})$$
$$= 11.85 \ \text{in}^2$$

Since all elements are connected, the shear lag coefficient, U, is 1.0.

$$A_e = U A_n$$
$$= (1.0)(11.85 \ \text{in}^2)$$
$$= 11.85 \ \text{in}^2$$

The axial tension is

$$P_n = A_g F_y$$
$$= (14.4 \ \text{in}^2)\left(50 \ \frac{\text{kips}}{\text{in}^2}\right)$$
$$= 720 \ \text{kips}$$
$$P_n = A_e F_u$$
$$= (11.85 \ \text{in}^2)\left(65 \ \frac{\text{kips}}{\text{in}^2}\right)$$
$$= 770 \ \text{kips}$$

For the ASD option, the allowable tensile strength is P_n/Ω_t.

For tensile yield, $\Omega_t = 1.67$.

$$\frac{P_n}{\Omega_t} = \frac{720 \ \text{kips}}{1.67} = 431 \ \text{kips}$$

For tensile rupture, $\Omega_t = 2.00$.

$$\frac{P_n}{\Omega_t} = \frac{770 \ \text{kips}}{2.00} = 385 \ \text{kips}$$

The controlling value is

$$\frac{P_n}{\Omega_t} = 385 \ \text{kips} \quad (390 \ \text{kips})$$

For the LRFD option, the design tensile strength is $\phi_t P_n$. For tensile yielding,

$$\phi_t = 0.90$$
$$\phi_t P_n = (0.90)(720 \ \text{kips})$$
$$= 648 \ \text{kips}$$

For tensile rupture,

$$\phi_t = 0.75$$
$$\phi_t P_n = (0.75)(770 \text{ kips})$$
$$= 578 \text{ kips}$$

The controlling value is

$$\phi_t P_n = 578 \text{ kips} \quad (580 \text{ kips})$$

The answer is (B).

156. Since the plates and channels do not experience differential movement relative to one another, AISC *Specification* Sec. E6 (built-up members) does not apply. The properties of a C12 × 30 are $A = 8.81 \text{ in}^2$, $t_w = 0.51 \text{ in}$, $I_x = 162 \text{ in}^4$ (strong axis), $I_y = 5.12 \text{ in}^4$ (weak axis), and the centroid is located 0.674 in from the outside edge of the web. For the built-up section,

$$I_x = \sum (I_{xc} + Ad^2)$$
$$= (2) \left(\frac{(0.5 \text{ in})(11 \text{ in})^3}{12} + 5.12 \text{ in}^4 \right. $$
$$\left. + (8.81 \text{ in}^2)\left(\frac{12 \text{ in}}{2} - 0.674 \text{ in}\right)^2 \right)$$
$$= 621.0 \text{ in}^4$$

$$I_y = \sum (I_{yc} + Ad^2)$$
$$= (2) \left(\frac{(11 \text{ in})(0.5 \text{ in})^3}{12} \right.$$
$$+ (11 \text{ in})(0.5 \text{ in})(6.25 \text{ in})^2$$
$$\left. + 162 \text{ in}^4 \right)$$
$$= 753.9 \text{ in}^4 \quad (754 \text{ in}^4)$$

$$A_g = \sum A_i$$
$$= (2) \Big((11 \text{ in})(0.5 \text{ in}) + 8.81 \text{ in}^2 \Big)$$
$$= 28.6 \text{ in}^2$$

$$r_x = \sqrt{\frac{I_x}{A}} = \sqrt{\frac{621.0 \text{ in}^4}{28.6 \text{ in}^2}}$$
$$= 4.66 \text{ in}$$

$$r_y = \sqrt{\frac{I_y}{A}} = \sqrt{\frac{754 \text{ in}^4}{28.6 \text{ in}^2}}$$
$$= 5.13 \text{ in}$$

The radius of gyration about the x-axis controls.

$$\frac{KL}{r} = \max \left\{ \frac{KL_x}{r_x}, \frac{KL_y}{r_y} \right\}$$
$$= \max \left\{ \frac{(22 \text{ ft})\left(12 \frac{\text{in}}{\text{ft}}\right)}{4.66 \text{ in}}, \frac{(24 \text{ ft})\left(12 \frac{\text{in}}{\text{ft}}\right)}{5.13 \text{ in}} \right\}$$
$$= \max \{57, 56\}$$
$$= 57$$

Check: $\quad \dfrac{KL}{r} \leq 4.71 \sqrt{\dfrac{E}{F_y}}$

$$57 \leq 4.71 \sqrt{\frac{29{,}000 \frac{\text{kips}}{\text{in}^2}}{36 \frac{\text{kips}}{\text{in}^2}}}$$

$$57 \leq 134 \quad [\text{OK}]$$

Calculate the critical stress.

$$F_{cr} = (0.658)^{F_y/F_e} F_y$$

$$F_e = \frac{\pi^2 E}{\left(\frac{KL}{r}\right)^2} = \frac{\pi^2 \left(29{,}000 \frac{\text{kips}}{\text{in}^2}\right)}{(56)^2}$$
$$= 91.3 \text{ ksi}$$

$$F_{cr} = (0.658)^{36 \frac{\text{kips}}{\text{in}^2}/91.3 \frac{\text{kips}}{\text{in}^2}} \left(36 \frac{\text{kips}}{\text{in}^2}\right)$$
$$= (0.658)^{0.39} \left(36 \frac{\text{kips}}{\text{in}^2}\right)$$
$$= 30.58 \text{ ksi}$$

$$P_n = F_{cr} A_g = \left(30.58 \frac{\text{kips}}{\text{in}^2}\right)(28.6 \text{ in}^2)$$
$$= 875 \text{ kips}$$

For the ASD option,

$$\frac{P_n}{\Omega} = \frac{875 \text{ kips}}{1.67}$$
$$= 524 \text{ kips} \quad (530 \text{ kips})$$

For the LRFD option,

$$\phi_c P_n = (0.9)(875 \text{ kips})$$
$$= 788 \text{ kips} \quad (790 \text{ kips})$$

The answer is (A).

157. For the 5.125 in × 22.5 in section,

$$S = \frac{bh^2}{6} = \frac{(5.125 \text{ in})(22.5 \text{ in})^2}{6}$$

$$= 432 \text{ in}^3$$

$$M_r = F_b' S$$

The allowable bending stress is obtained by multiplying the basic stress by the lesser of the volume factor, C_v, and the lateral stability factor, C_L.

$$C_v = \left(\left(\frac{21}{L} \right) \left(\frac{12}{d} \right) \left(\frac{5.125}{b} \right) \right)^{1/x}$$

$$= \left(\left(\frac{21 \text{ ft}}{32 \text{ ft}} \right) \left(\frac{12 \text{ in}}{22.5 \text{ in}} \right) \left(\frac{5.125 \text{ in}}{5.125 \text{ in}} \right) \right)^{1/10}$$

$$= 0.90$$

$$L_u = \frac{32 \text{ ft}}{2} = 16 \text{ ft}$$

$$\frac{L_u}{d} = \frac{(16 \text{ ft})\left(12 \, \dfrac{\text{in}}{\text{ft}}\right)}{22.5 \text{ in}}$$

$$= 8.5$$

$$7 < \frac{L_u}{d} < 14.3$$

Therefore,

$$L_e = 1.63 L_u + 3d = (1.63)(16 \text{ ft})\left(12 \, \frac{\text{in}}{\text{ft}}\right) + (3)(22.5 \text{ in})$$

$$= 380.5 \text{ in} \quad (381 \text{ in})$$

$$R_B = \sqrt{\frac{L_e d}{b^2}} = \sqrt{\frac{(381 \text{ in})(22.5 \text{ in})}{(5.125 \text{ in})^2}}$$

$$= 18$$

$$F_b^* = F_b = 2400 \text{ lbf/in}^2$$

$$F_{bE} = \frac{1.20 E_{\min}'}{R_B^2} = \frac{(1.20)\left(830{,}000 \, \dfrac{\text{lbf}}{\text{in}^2}\right)}{(18)^2}$$

$$= 3074 \text{ lbf/in}^2$$

$$\frac{F_{bE}}{F_b^*} = \frac{3074 \, \dfrac{\text{lbf}}{\text{in}^2}}{2400 \, \dfrac{\text{lbf}}{\text{in}^2}} = 1.28$$

$$C_L = \frac{1 + \dfrac{F_{bE}}{F_b^*}}{1.9} - \sqrt{\left(\frac{1 + \dfrac{F_{bE}}{F_b^*}}{1.9}\right)^2 - \frac{\dfrac{F_{bE}}{F_b^*}}{0.95}}$$

$$= \frac{1 + 1.26}{1.9} - \sqrt{\left(\frac{1 + 1.26}{1.9}\right)^2 - \frac{1.26}{0.95}}$$

$$= 0.89$$

Allowable bending stress is controlled by the stability factor, C_L.

$$M_r = F_b' S = C_L F_b S = \frac{(0.89)\left(2400 \, \dfrac{\text{lbf}}{\text{in}^2}\right)(432 \text{ in}^3)}{\left(1000 \, \dfrac{\text{lbf}}{\text{kip}}\right)\left(12 \, \dfrac{\text{in}}{\text{ft}}\right)}$$

$$= 77 \text{ ft-kips} \quad (80 \text{ ft-kips})$$

The answer is (C).

158.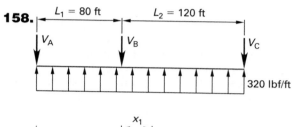

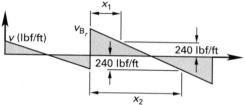

For a plywood diaphragm, the reactions to the shear walls are in proportion to their tributary lengths.

The shear in the diaphragm just to the right of shear wall B is

$$V_{B,r} = \frac{wL_2}{2} = \frac{\left(320 \, \dfrac{\text{lbf}}{\text{ft}}\right)(120 \text{ ft})}{2}$$

$$= 19{,}200 \text{ lbf}$$

The shear distributes uniformly over the width of the diaphragm.

$$v_{B,r} = \frac{V_{B,r}}{B} = \frac{19{,}200 \text{ lbf}}{40 \text{ ft}}$$

$$= 480 \text{ lbf/ft}$$

The variation in diaphragm shear between B and C is

$$v = v_{B,r} - mx$$

$$= 480 \, \frac{\text{lbf}}{\text{ft}} - \left(\frac{960 \, \dfrac{\text{lbf}}{\text{ft}}}{120 \text{ ft}}\right)x$$

The limits of the region where the absolute value of v is less than 240 lbf/ft are given by the following equation.

$$v = 240 \, \frac{\text{lbf}}{\text{ft}} = \left| 480 \, \frac{\text{lbf}}{\text{ft}} - \left(8 \, \frac{\text{lbf}}{\text{ft}^2}\right)x \right|$$

$$x_1 = 30 \text{ ft}$$

$$x_2 = 90 \text{ ft}$$

The answer is (C).

159. The number of loading cycles is

$$N = \left(50 \ \frac{\text{cycles}}{\text{day}}\right)\left(365 \ \frac{\text{days}}{\text{yr}}\right)(25 \text{ yr design life})$$
$$= 456{,}250 \text{ cycles}$$

Per the *AISC Manual* App. 3 Eq. A-3-1, the design stress range is determined by

$$F_{\text{SR}} = \left(\frac{C_f}{N}\right)^{0.333} \geq F_{\text{TH}}$$

The bottom flange will experience alternating tension stress adjacent to the transverse weld, which corresponds to stress category E'. (The flange thickness of the W21 × 93 is 0.93 in, which exceeds the limit of 0.8 in that distinguishes categories E and E'.) For stress category E', constant C_f is 3.9×10^8, and the threshold fatigue stress range, F_{TH}, is 2.6 ksi.

$$F_{\text{SR}} = \left(\frac{C_f}{N}\right)^{0.333}$$
$$= \left(\frac{3.9 \times 10^8}{456{,}250 \text{ cycles}}\right)^{0.333}$$
$$= 9.5 \text{ ksi} \quad [\geq 2.6 \text{ ksi}]$$

The answer is (A).

160. Design on a unit-length basis ($b = 12$ in).

$$M_{\text{max}} = \frac{wH^2}{8} = \frac{\left(20 \ \frac{\text{lbf}}{\text{ft}}\right)(15.33 \text{ ft})^2}{8}$$
$$= 587.8 \text{ ft-lbf} \quad (588 \text{ ft-lbf})$$

For wall reinforcement centered in a nominal 8 in masonry wall,

$$d = \frac{t}{2} = \frac{7.62 \text{ in}}{2}$$
$$= 3.81 \text{ in}$$

For grade 60 reinforcement subject to wind load,

$$F_s = 24{,}000 \text{ lbf/in}^2$$

For the masonry,

$$F_c = \frac{f'_m}{3} = \frac{1500 \ \frac{\text{lbf}}{\text{in}^2}}{3}$$
$$= 500 \text{ lbf/in}^2$$

A trial value of A_s is found using an internal lever arm of $^7/_8$ of the effective depth.

$$A_s \approx \frac{M_{\text{max}}}{F_s \frac{7}{8} d}$$
$$= \frac{(588 \text{ ft-lbf})\left(12 \ \frac{\text{in}}{\text{ft}}\right)}{\left(24{,}000 \ \frac{\text{lbf}}{\text{in}^2}\right)\left(\frac{7}{8}\right)(3.81 \text{ in})}$$
$$= 0.088 \text{ in}^2$$

The steel ratio is

$$\rho = \frac{A_s}{bd} = \frac{0.088 \text{ in}^2}{(12 \text{ in})(3.81 \text{ in})}$$
$$= 0.001929$$

From ACI 530 Sec. 1.8.2.2.1, $E_m = 900 f'_m$. The modular ratio is

$$n = \frac{E_s}{E_m} = \frac{E_s}{900 f'_m}$$
$$= \frac{29{,}000{,}000 \ \frac{\text{lbf}}{\text{in}^2}}{(900)\left(1500 \ \frac{\text{lbf}}{\text{in}^2}\right)}$$
$$= 21.48$$
$$\rho n = (0.001929)(21.48)$$
$$= 0.0414$$

The depth to the neutral axis is

$$k = \sqrt{(\rho n)^2 + 2\rho n} - \rho n$$
$$= \sqrt{(0.0414)^2 + (2)(0.0414)} - 0.0414$$
$$= 0.249$$

$$j = 1 - \frac{k}{3}$$
$$= 1 - \frac{0.249}{3}$$
$$= 0.917$$

The stress in the steel is

$$f_s = \frac{M_{\text{max}}}{A_s j d}$$
$$= \frac{(588 \text{ ft-lbf})\left(12 \ \frac{\text{in}}{\text{ft}}\right)}{(0.088 \text{ in}^2)(0.917)(3.81 \text{ in})}$$
$$= 22{,}900 \text{ lbf/in}^2 < F_s$$

The stress in concrete is

$$f_c = \left(\frac{M_{max}}{bd^2}\right)\left(\frac{2}{jk}\right)$$

$$= \left(\frac{(588 \text{ ft-lbf})\left(12 \frac{\text{in}}{\text{ft}}\right)}{(12 \text{ in})(3.81 \text{ in})^2}\right)\left(\frac{2}{(0.917)(0.249)}\right)$$

$$= 354.3 \text{ lbf/in}^2 < F_c$$

The trial value of $A_s = 0.088 \text{ in}^2$ (0.09 in^2) is sufficient.

The answer is (B).

Solutions
Afternoon Session Transportation

161. The curve radius is

$$R = \frac{(360°)(100 \text{ ft})}{2\pi D} = \frac{(360°)(100 \text{ ft})}{2\pi(2°)}$$
$$= 2864.789 \text{ ft}$$

The interior angle is

$$I = 54° \, 56' \, 24'' - 32° \, 15' \, 18''$$
$$= 22° \, 41' \, 6'' \quad (22.685°)$$

The tangent length between PI and PT is

$$T = R \tan \frac{I}{2} = (2864.789 \text{ ft}) \tan\left(\frac{22.685°}{2}\right)$$
$$= 574.651 \text{ ft}$$

The northing (i.e., change in north dimension) is

$$\Delta N = T \cos(\text{bearing})$$
$$= (574.651 \text{ ft}) \cos(54° \, 56' \, 24'')$$
$$= 330.099 \text{ ft}$$

The easting (i.e., change in east dimension) is

$$\Delta E = T \sin(\text{bearing})$$
$$= (574.651 \text{ ft}) \sin(54° \, 56' \, 24'')$$
$$= 470.382 \text{ ft}$$

The northing of the PT is

$$N_{PT} = N_{PI} + \Delta N$$
$$= 423{,}968.68 \text{ ft} + 330.10 \text{ ft}$$
$$= 424{,}298.78 \text{ ft}$$

The easting of the PT is

$$E_{PT} = E_{PI} + \Delta E$$
$$= 268{,}236.42 \text{ ft} + 470.38 \text{ ft}$$
$$= 268{,}706.80 \text{ ft}$$

The coordinates of the PT are 424,298.78 N and 268,706.80 E.

The answer is (B).

162. Find the LOS using *HCM* Exh. 11-17. At the given free-flow speed of 65 mph, the maximum service flow rate (1350 pcphpl) is greater than the maximum service flow rate for LOS B (1170 pcphpl) and less than the maximum service flow rate for LOS C (1630 pcphpl). Therefore, the freeway is considered to be operating at LOS C.

The answer is (B).

163. First determine the density, D, from the given equation using the given average travel speed of 50 mph.

$$S = 65 \, \frac{\text{mi}}{\text{hr}} - \left(0.42 \, \frac{\frac{\text{mi}}{\text{hr}}}{\frac{\text{veh}}{\text{mi}}}\right) D$$

$$50 \, \frac{\text{mi}}{\text{hr}} = 65 \, \frac{\text{mi}}{\text{hr}} - \left(0.42 \, \frac{\frac{\text{mi}}{\text{hr}}}{\frac{\text{veh}}{\text{mi}}}\right) D$$

$$D = 35.7 \text{ vpm}$$

The flow rate can be calculated using the *HCM* traffic density equation (*HCM* Eq. 4-4), given the speed and density.

$$D = \frac{v}{S}$$
$$35.7 \, \frac{\text{veh}}{\text{mi}} = \frac{v}{50 \, \frac{\text{mi}}{\text{hr}}}$$
$$v = 1786 \text{ vph} \quad (1800 \text{ vph})$$

The answer is (B).

164. Use the *AASHTO Green Book*. For rural highways, assume the maximum rate of superelevation, 0.08–0.10 ft/ft. For 40 mph design speed and a 2° curve, the curve radius is

$$R = \frac{(360°)(100 \text{ ft})}{2\pi D}$$
$$= \frac{(360°)(100 \text{ ft})}{2\pi(2°)}$$
$$= 2864.79 \text{ ft}$$

In *Green Book* Fig. 3-11 and Fig. 3-12, the resulting rate of superelevation is approximately 2.8% (8%) to 2.9% (10%) (0.03 ft/ft).

The answer is (B).

165. The curve length is the distance from the PC to the PT.

$$L = \frac{2\pi R I}{360°} = \frac{2\pi(2080 \text{ ft})(60°)}{360°}$$
$$= 2178.17 \text{ ft}$$
$$\text{sta PT} = \text{sta PC} + L = (\text{sta } 12{+}40) + 2178.17 \text{ ft}$$
$$= \text{sta } 34{+}18.17 \quad (\text{sta } 34{+}18)$$

The answer is (C).

166. Use *AASHTO Green Book* Table 3-1 on stopping sight distance. The minimum stopping sight distance for a design speed of 40 mph is 300.6 ft. (The design value is rounded up to 305 ft.)

The answer is (C).

167. From the given table, the runoff for a curve of 2080 ft radius is 150 ft. On circular curves such as this, the 150 ft represents one-third of the superelevation runoff, L (the transition from normal cross slope to the superelevated cross section). Therefore, two-thirds of the superelevation runoff is developed on the tangent, and in this case, that is 300 ft. The tangent runout, T_R (the distance until the adverse cross slope has been removed), is 300 ft (twice the runoff in the curve according to the problem statement). The station of the transition start (TS) is

$$\text{sta TS} = \text{sta PC} - \tfrac{2}{3}L - T_R$$
$$= (\text{sta } 12{+}40) - 300 \text{ ft} - 300 \text{ ft}$$
$$= \text{sta } 6{+}40$$

The answer is (A).

168. The midpoint, M, of the curve is located half of the curve length, L, past the PC.

$$\text{sta M} = \text{sta PC} + \frac{L}{2}$$
$$= (\text{sta } 13{+}50) + \frac{2000 \text{ ft}}{2}$$
$$= \text{sta } 23{+}50$$

The elevation of the centerline at M is

$$\text{elev}_{\text{M,centerline}} = \text{elev}_{\text{PC}} + G\left(\frac{L}{2}\right)$$
$$= 170 \text{ ft} + \left(0.0075 \ \frac{\text{ft}}{\text{ft}}\right)\left(\frac{2000 \text{ ft}}{2}\right)$$
$$= 177.5 \text{ ft}$$

Since the midpoint is more than 150 ft from the PC and more than 150 ft from the PT, the section is fully elevated at the midpoint. The cross slope is 0.045 ft/ft. The outside pavement edge is higher than the centerline by

$$\Delta_{\text{elev}} = \left(0.045 \ \frac{\text{ft}}{\text{ft}}\right)\left(12 \ \frac{\text{ft}}{\text{lane}}\right)(1 \text{ lane})$$
$$= 0.54 \text{ ft}$$

The elevation of the outside pavement edge is

$$\text{elev}_{\text{M,edge}} = \text{elev}_{\text{M,centerline}} + \Delta_{\text{elev}}$$
$$= 177.5 \text{ ft} + 0.54 \text{ ft}$$
$$= 178.04 \text{ ft} \quad (178 \text{ ft})$$

The answer is (D).

169. The peak hour factor (PHF) is defined in the *HCM* as the ratio of the total hourly volume to the peak rate of flow within the hour.

$$\text{PHF} = \frac{\text{hourly volume}}{\text{peak rate of flow within the hour}}$$
$$= \frac{V}{4V_{15}}$$
$$= \frac{195 \text{ veh} + 163 \text{ veh} + 157 \text{ veh} + 178 \text{ veh}}{(4)(195 \text{ veh})}$$
$$= 0.888 \quad (0.89)$$

The answer is (B).

170. The recreational vehicles can be combined with trucks because the percentage of trucks and buses (22%) is at least five times the percentage of RVs (4%) (*HCM* p. 11-14). Therefore, the analysis will be based on a percentage of trucks, P_T, of 26%.

Since some grades exceed 2%, classify the terrain as rolling. From *HCM* Exh. 11-11, the passenger car equivalent of heavy trucks, E_T, is 1.5. The heavy vehicle adjustment factor, f_{HV}, is determined using *HCM* Eq. 11-3.

$$f_{HV} = \frac{1}{1 + P_T(E_T - 1)}$$
$$= \frac{1}{1 + (0.26)(1.5 - 1)}$$
$$= 0.885$$

The driver population factor, f_p, is assumed to be 1.0 for commuter traffic. The 15 min passenger car equivalent flow rate is

$$v_p = \frac{V}{(PHF)Nf_{HV}f_p}$$
$$= \frac{2500 \frac{veh}{hr}}{(0.85)(3 \text{ lanes})(0.885)(1.0)}$$
$$= 1108 \text{ pcphpl} \quad (1110 \text{ pcphpl})$$

The answer is (B).

171. The actual free-flow speed (FFS) can be determined from *HCM* Eq. 11-1.

$$f_{LW} = 1.9 \text{ mph} \quad [HCM \text{ Exh. 11-8}]$$
$$f_{LC} = 0.80 \quad [HCM \text{ Exh. 11-9}]$$
$$TRD = 1 \text{ ramp}/2 \text{ mi} = 0.5 \text{ ramp/mi}$$
$$FFS = 75.4 \text{ mph} - f_{LW} - f_{LC} - 3.22TRD^{0.84}$$
$$= 75.4 \text{ mph} - 1.9 \text{ mph} - 0.80 \text{ mph}$$
$$- (3.22)\left(0.5 \frac{ramp}{mi}\right)^{0.84}$$
$$= 70.9 \text{ mph} \quad (71 \text{ mph})$$

The FFS is used as the average passenger car speed, S. The density of flow from *HCM* Eq. 11-4 is

$$D = \frac{v_p}{S} = \frac{1938 \text{ pcphpl}}{71 \frac{mi}{hr}}$$
$$= 27.3 \text{ pcpmpl}$$

From *HCM* Exh. 11-5, the LOS is D.

The answer is (C).

172. Assuming three lanes in each direction, the free-flow speed (FFS) can be determined from *HCM* Eq. 11-1.

$$f_{LW} = 0.0 \text{ mph} \quad [HCM \text{ Exh. 11-8}]$$
$$f_{LC} = 0.0 \text{ mph} \quad [HCM \text{ Exh. 11-9}]$$
$$TRD = 1 \text{ ramp/mi}$$
$$FFS = 75.4 \text{ mph} - f_{LW} - f_{LC} - 3.22TRD^{0.84}$$
$$= 75.4 \text{ mph} - 0.0 \text{ mph} - 0.0 \text{ mph}$$
$$- (3.22)\left(1 \frac{ramp}{mi}\right)^{0.84}$$
$$= 72.18 \text{ mph}$$

The FFS is used as the average passenger car speed, S. The density of flow for a six-lane freeway (three lanes in each direction) is

$$D_{3 \text{ lanes}} = \frac{v_{p3 \text{ lanes}}}{S_{3 \text{ lanes}}} = \frac{1910 \text{ pcphpl}}{72.18 \frac{mi}{hr}}$$
$$= 26.46 \text{ pcpmpl}$$

The calculated density (26.46 pcpmpl) is greater than 26 pcpmpl (the maximum density for LOS C from *HCM* Exh. 11-5). The expected LOS is D if only three lanes are provided in each direction. Therefore, revise the proposed design and provide four lanes in each direction and check the density again.

The change in number of lanes will also change the 15 min passenger car equivalent flow rate ($v_p = 1910$ pcphpl) because it was based on a six-lane freeway.

$$v_{p,4 \text{ lanes}} = \tfrac{3}{4} v_{p,3 \text{ lanes}}$$
$$= \left(\frac{3}{4}\right)(1910 \text{ pcphpl})$$
$$= 1433 \text{ pcphpl}$$

Assume four lanes in each direction. The density of flow for an eight-lane freeway (four lanes in each direction) is given by *HCM* Eq. 11-4.

$$D_{4 \text{ lanes}} = \frac{v_{p,4 \text{ lanes}}}{S_{4 \text{ lanes}}} = \frac{1433 \text{ pcphpl}}{72.18 \frac{mi}{hr}}$$
$$= 19.85 \text{ pcpmpl}$$

The calculated density (19.85 pcpmpl) is less than 26 pcpmpl (the maximum density for LOS C from *HCM* Exh. 11-5). Therefore, the expected freeway LOS is C if four lanes are provided in each direction.

The answer is (C).

173. Use *HCM* Exh. 11-17. The freeway capacity per lane when $v/c = 1.0$ is 2300 pcphpl at 60 mph free-flow speed. For a growth factor of $i = 5\%$, the number of years, n, until the freeway reaches capacity can be determined as follows.

$$v_{p,\text{future}} = v_{p,\text{present}}(1 + i)^n$$

$$2300 \text{ pcphpl} = (1900 \text{ pcphpl})(1 + 0.05)^n$$

$$\log 2300 \text{ pcphpl} = \log 1900 \text{ pcphpl} + n \log(1 + 0.05)$$

$$n = 3.9 \text{ yr} \quad (4 \text{ yr})$$

The answer is (D).

174. The horizontal distance from the BVC track centerline is

$$x_{\text{rail}} = \frac{L}{2} + (\text{sta } 28{+}50) - (\text{sta } 26{+}00)$$

$$= \frac{16.48 \text{ sta}}{2} + (\text{sta } 28{+}50) - (\text{sta } 26{+}00)$$

$$= 10.74 \text{ sta}$$

$$\text{elev}_{\text{BVC}} = \text{elev}_{\text{V}} - G_1\left(\frac{L}{2}\right)$$

$$= 231.00 \text{ ft} - \left(3 \frac{\text{ft}}{\text{sta}}\right)\left(\frac{16.48 \text{ sta}}{2}\right)$$

$$= 206.28 \text{ ft}$$

The curve elevation at track centerline is

$$\text{elev}_{28{+}50} = \frac{(G_2 - G_1)x_{\text{rail}}^2}{2L} + G_1 x_{\text{rail}} + \text{elev}_{\text{BVC}}$$

$$= \frac{\left(-2 \frac{\text{ft}}{\text{sta}} - 3 \frac{\text{ft}}{\text{sta}}\right)(10.74 \text{ sta})^2}{(2)(16.48 \text{ sta})}$$

$$+ \left(3 \frac{\text{ft}}{\text{sta}}\right)(10.74 \text{ sta}) + 206.28 \text{ ft}$$

$$= 221.00 \text{ ft}$$

The distance, D, is the difference between the curve elevation at the track centerline and the elevation of the railbed at that same point.

$$D = 221.00 \text{ ft} - 195.00 \text{ ft}$$

$$= 26 \text{ ft}$$

The answer is (C).

175. There are different ways to determine the minimum length of a vertical curve. The quickest is to use *AASHTO Green Book* Fig. 3-43, design controls for crest vertical curves with open road conditions.

Calculate the algebraic difference, A, in grades.

$$A = |G_2 - G_1|$$

$$= |-6\% - 2\%|$$

$$= 8\%$$

Enter a horizontal line from the algebraic difference in grades into *Green Book* Fig. 3-43 to intersect the 55 mph curve. Drop a vertical line from the point of intersection, and read the minimum length of curve, L, which is approximately 900 ft.

The answer is (D).

176. The elevation of the EVC is

$$\text{elev}_{\text{EVC}} = \text{elev}_{\text{V}} + G_2\left(\frac{L}{2}\right)$$

$$= 231.00 \text{ ft} + \left(-2 \frac{\text{ft}}{\text{sta}}\right)\left(\frac{16.48 \text{ sta}}{2}\right)$$

$$= 214.52 \text{ ft} \quad (210 \text{ ft})$$

The answer is (C).

177. The horizontal distance from the BVC is

$$x_{\text{high}} = \frac{LG_1}{|G_2 - G_1|}$$

$$= \frac{(22.00 \text{ sta})\left(3 \frac{\text{ft}}{\text{sta}}\right)}{\left|-5 \frac{\text{ft}}{\text{sta}} - 3 \frac{\text{ft}}{\text{sta}}\right|}$$

$$= 8.25 \text{ sta}$$

The station of the BVC is

$$\text{sta}_{\text{BVC}} = \text{sta}_{\text{V}} - \frac{L}{2}$$

$$= (\text{sta } 91{+}70) - \frac{22.00 \text{ sta}}{2}$$

$$= \text{sta } 80{+}70$$

The station of the highest point on the curve (the turning point) is

$$\text{sta}_{\text{high}} = \text{sta}_{\text{BVC}} + x_{\text{high}}$$

$$= (\text{sta } 80{+}70) + 8.25 \text{ sta}$$

$$= \text{sta } 88{+}95$$

The answer is (B).

178. The distance from the point of intersection of the two tangents (the vertex) to the midpoint of the curve is the external distance, E.

$$E = R\left(\sec\frac{I}{2} - 1\right)$$

$$= (1100 \text{ ft})\left(\sec\frac{45°}{2} - 1\right)$$

$$= 90.63 \text{ ft} \quad (91 \text{ ft})$$

The answer is (C).

179. The horizontal distance from the BVC is

$$x_{\text{low}} = \frac{-LG_1}{|G_2 - G_1|} = \frac{-(9 \text{ sta})\left(-4\ \dfrac{\text{ft}}{\text{sta}}\right)}{\left|1\ \dfrac{\text{ft}}{\text{sta}} - \left(-4\ \dfrac{\text{ft}}{\text{sta}}\right)\right|} = 7.20 \text{ sta}$$

The elevation of the BVC is

$$\text{elev}_{\text{BVC}} = \text{elev}_{\text{V}} - G_1\left(\frac{L}{2}\right)$$

$$= 2231.31 \text{ ft} - \left(-4\ \frac{\text{ft}}{\text{sta}}\right)\left(\frac{9.00 \text{ sta}}{2}\right)$$

$$= 2249.31 \text{ ft}$$

The elevation of the lowest point on the curve (the turning point) is

$$\text{elev}_{\text{low}} = \frac{(G_2 - G_1)x_{\text{low}}^2}{2L} + G_1 x_{\text{low}} + \text{elev}_{\text{BVC}}$$

$$= \frac{\left(1\ \dfrac{\text{ft}}{\text{sta}} - \left(-4\ \dfrac{\text{ft}}{\text{sta}}\right)\right)(7.20 \text{ sta})^2}{(2)(9.00 \text{ sta})}$$

$$\quad + \left(-4\ \frac{\text{ft}}{\text{sta}}\right)(7.20 \text{ sta}) + 2249.31 \text{ ft}$$

$$= 2234.91 \text{ ft} \quad (2235 \text{ ft})$$

The answer is (D).

180. Use the procedure for analysis of two-lane highways with general terrain since no specific grade was given.

The free-flow speed can be estimated using *HCM* Eq. 15-2.

$$f_{\text{LS}} = 2.4 \text{ mph} \quad [\textit{HCM} \text{ Exh. 15-7}]$$

$$f_A = 0.5 \text{ mph} \quad [\textit{HCM} \text{ Exh. 15-8, via interpolation}]$$

$$\text{FFS} = \text{BFFS} - f_{\text{LS}} - f_A$$

$$= 60 \text{ mph} - 2.4 \text{ mph} - 0.5 \text{ mph}$$

$$= 57.1 \text{ mph}$$

The question asks for a directional peak flow. The unadjusted one-way traffic is

$$(0.6)(1100 \text{ pcph}) = 660 \text{ pcph}$$

The opposing demand flow rate is

$$(0.4)(1100 \text{ pcph}) = 440 \text{ pcph}$$

The highest directional flow rate is based on the average travel speed (ATS) or the percent time spent following (PTSF). The directional flow rate based on ATS in the analysis direction is found using *HCM* Eq. 15-3.

$$v_{i,\text{ATS}} = \frac{V}{(\text{PHF})f_{g,\text{ATS}}f_{\text{HV,ATS}}}$$

$$= \frac{660 \text{ pcph}}{(0.92)(1.0)(0.909)}$$

$$= 789 \text{ pcph}$$

Find the directional flow rate based on the PTSF. The heavy vehicle adjustment factor for PTSF is found using *HCM* Eq. 15-8 and Exh. 15-18 for passenger car equivalents for trucks and RVs.

$$f_{\text{HV,PTSF}} = \frac{1}{1 + P_T(E_T - 1) + P_R(E_R - 1)}$$

$$= \frac{1}{1 + (0.08)(1.0 - 1) + (0.02)(1.0 - 1)}$$

$$= 1.0$$

$f_{g,\text{PTSF}} = 1.0$.

The directional flow rate based on PTSF is found using *HCM* Eq. 15-7 and Exh. 15-18 for passenger car equivalents for trucks and RVs. The PTSF demand in the analysis direction is

$$v_{d,\text{PTSF}} = \frac{V_d}{(\text{PHF})f_{g,\text{PTSF}}f_{\text{HV,PTSF}}}$$

$$= \frac{660\ \dfrac{\text{veh}}{\text{hr}}}{(0.92)(1.0)(1.0)}$$

$$= 717 \text{ vph}$$

Comparing ATS and PTSF, the highest directional flow rate for the peak 15 min period is 789 vph (790 vph).

The answer is (B).

181. Estimate the expected traffic volume that will use the new freeway by plotting the point of intersection of the demand and supply curves. The two curves can be combined as shown in *Illustration for Solution 181*. At the point of intersection, the volume is most nearly 1750 vph.

The answer is (D).

182. Total overhaul distance (OHD) is the difference between the average total haul distance (THD) and free haul distance (FHD).

$$\text{OHD} = \text{THD} - \text{FHD}$$
$$= 840 \text{ ft} - 500 \text{ ft}$$
$$= 340 \text{ ft}$$

Assuming 100 ft stations, the OHD in stations is

$$\text{OHD}_s = \frac{340 \text{ ft}}{100 \ \frac{\text{sta}}{\text{ft}}} = 3.40 \text{ sta}$$

The overhaul in cubic yards (OH_v) is 730 yd^3 from the mass diagram for 500 ft free haul. The overhaul in cubic yard-stations (OH) is

$$\text{OH} = (\text{OH}_v)(\text{OHD}_s)$$
$$= (730 \text{ yd}^3)(3.40 \text{ sta})$$
$$= 2482 \text{ yd}^3\text{-sta}$$

The overhaul cost (OHC) depends on the overhaul unit cost (UC) and the amount of overhaul.

$$\text{OHC} = (\text{UC})(\text{OH})$$
$$= \left(9.75 \ \frac{\$}{\text{yd}^3\text{-sta}}\right)(2482 \text{ yd}^3\text{-sta})$$
$$= \$24{,}200 \quad (\$24{,}000)$$

The answer is (C).

183. The service rate per hour, Q, depends on the service time per vehicle (SPV).

$$Q = \frac{1}{\text{SPV}}$$
$$= \left(\frac{1}{1.5 \ \frac{\text{min}}{\text{veh}}}\right)\left(60 \ \frac{\text{min}}{\text{hr}}\right)$$
$$= 40 \text{ vph}$$

The service rate is greater than the arrival rate, q, of 30 vph. Therefore, the queue is undersaturated. The expected number of vehicles waiting in the queue is

$$L_q = \frac{q^2}{Q(Q - q)}$$
$$= \frac{\left(30 \ \frac{\text{veh}}{\text{hr}}\right)^2}{\left(40 \ \frac{\text{veh}}{\text{hr}}\right)\left(40 \ \frac{\text{veh}}{\text{hr}} - 30 \ \frac{\text{veh}}{\text{hr}}\right)}$$
$$= 2.25 \quad (3)$$

The answer is (B).

184. The expected number of fatal accidents after the development, FA_a, is the number of fatal accidents before the development, FA_b, reduced by 25%.

$$\text{FA}_a = (1 - \text{reduction})\text{FA}_b$$
$$= (1 - 0.25)(140 \text{ total accidents})$$
$$\times \left(\frac{5 \text{ fatal accidents}}{100 \text{ total accidents}}\right)$$
$$= 5.25 \text{ fatal accidents}$$

Illustration for Solution 181

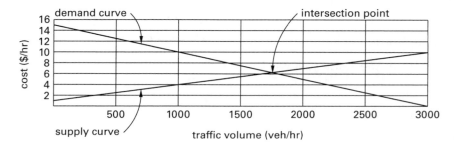

The vehicle-miles of travel (VMT) during the 3 yr period, T, after the development and improvements on this 15 mile section, L, is

$$\text{VMT} = (\text{ADT})\,T L$$

$$= \left(18{,}000\ \frac{\text{veh}}{\text{day}}\right)(3\ \text{yr})\left(365\ \frac{\text{day}}{\text{yr}}\right)(15\ \text{mi})$$

$$= 295.65 \times 10^6\ \text{veh-mi}$$

The fatal accident rate, RPVM, per 100 million vehicle-miles of travel (HMVM) is

$$\text{RPVM} = \frac{\text{FA}_a(100 \times 10^6)}{\text{VMT}}$$

$$= \frac{(5.25\ \text{fatal accidents})(100 \times 10^6)}{295.65 \times 10^6\ \text{veh-mi}}$$

$$= 1.78\ \text{fatal accidents/HMVM}$$

$$(1.8\ \text{fatal accidents/HMVM})$$

The answer is (B).

---

185. Using the gravity model and the productions, P, and attractions, A, of the related zones, the trips produced in zone 1 and attracted to zone 3 are

$$T_{1,3} = P_1\left(\frac{A_3 F_{1,3} K_{1,3}}{A_2 F_{1,2} K_{1,2} + A_3 F_{1,3} K_{1,3} + A_4 F_{1,4} K_{1,4}}\right)$$

$$= (1600\ \text{trips})\left(\frac{(1100\ \text{trips})(8)(1)}{\begin{array}{l}(840\ \text{trips})(10)(1) \\[4pt] + (1100\ \text{trips})(8)(1) \\[4pt] + (650\ \text{trips})(24)(1)\end{array}}\right)$$

$$= 429.3\ \text{trips} \quad (430\ \text{trips})$$

The factor K is assumed to be 1 because the problem stated that the socioeconomic conditions are the same for all zones.

The answer is (B).

---

186. Based on the node-to-node travel times from the illustration in the problem, the minimum travel times from node 1 to all the other nodes are shown in the following illustration. The links with the longer total travel times from node 1 are eliminated based on an all-or-nothing approach.

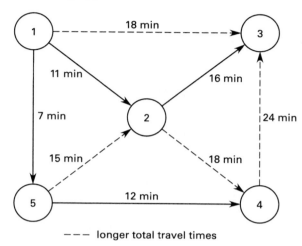

--- longer total travel times

(not to scale)

Link 1-5 will carry the volume from node 1 to both nodes 5 and 4. The total volume, V, on link 1-5 is 245 veh (115 veh + 130 veh) as shown in the following illustration.

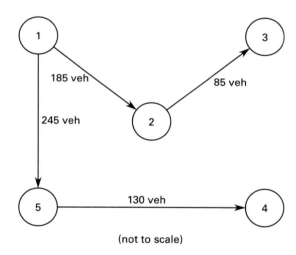

(not to scale)

The vehicle minutes of travel, VM, is the total volume, V, multiplied by the travel time, T.

$$\text{VM} = V T = (245\ \text{veh})(7\ \text{min})$$

$$= 1715\ \text{veh-min} \quad (1700\ \text{veh-min})$$

The answer is (C).

187. The QRS method is based on incorporating service elements in the estimation. Impedance (also known as utility propensity, friction factor, or probability function) of the auto mode between zones 1 and 2, $I_{1,2a}$, is a function of the in-vehicle time, VT_a, in minutes; the excess time, ET_a, in minutes; the trip cost, TC_a; and the income per min, IN_a.

$$
\begin{aligned}
I_{1,2a} &= VT_a + 2.5ET_a + 3\,\frac{TC_a}{IN_a} \\
&= \left(\frac{12\ \text{mi}}{55\ \frac{\text{mi}}{\text{hr}}}\right)\left(60\ \frac{\text{min}}{\text{hr}}\right) + (2.5)(6\ \text{min}) \\
&\quad + (3)\left(\frac{\left(0.35\ \frac{\$}{\text{mi}}\right)(12\ \text{mi})}{\dfrac{\$30{,}000}{120{,}000\ \text{min}}}\right) \\
&= 78.49\ \text{min}
\end{aligned}
$$

Similarly, impedance of the transit mode between zones 1 and 2 is

$$
\begin{aligned}
I_{1,2t} &= VT_t + 2.5ET_t + 3\,\frac{TC_t}{IN_t} \\
&= \left(\frac{10\ \text{mi}}{45\ \frac{\text{mi}}{\text{hr}}}\right)\left(60\ \frac{\text{min}}{\text{hr}}\right) + (2.5)(11\ \text{min}) \\
&\quad + (3)\left(\frac{\left(0.20\ \frac{\$}{\text{mi}}\right)(10\ \text{mi})}{\dfrac{\$30{,}000}{120{,}000\ \text{min}}}\right) \\
&= 64.83\ \text{min}
\end{aligned}
$$

Since the impedance is a cost and time function, higher impedances will be disincentives to use these modes of travel. The percentage of trips between zones 1 and 2 that are expected to use auto, MS_a, is specified by the QRS model as proportional to the transit impedance.

$$
\begin{aligned}
MS_a &= \frac{I_{1,2t}^b}{I_{1,2t}^b + I_{1,2a}^b} \times 100\% \\
&= \frac{(64.83\ \text{min})^{1.5}}{(64.83\ \text{min})^{1.5} + (78.49\ \text{min})^{1.5}} \times 100\% \\
&= 42.9\% \quad (43\%)
\end{aligned}
$$

The answer is (A).

188. The total number of vehicles entering the intersections is

$$
\begin{aligned}
VE &= \sum \text{entering traffic} \\
&= 1250\ \frac{\text{veh}}{\text{day}} + 2350\ \frac{\text{veh}}{\text{day}} + 730\ \frac{\text{veh}}{\text{day}} + 1920\ \frac{\text{veh}}{\text{day}} \\
&= 6250\ \text{veh/day}
\end{aligned}
$$

Accident rates are reported separately for fatal and incapacitating accidents, other injury accidents, and accidents with property damage only. This problem asks for the (non-fatal) injury accident rate. From the problem statement, 13 injury-causing accidents, IA, occurred in 1 yr. The rate of IA per 10 million entering vehicles is

$$
\begin{aligned}
RPEV &= \frac{IA(10 \times 10^6)}{VE\left(365\ \frac{\text{day}}{\text{yr}}\right)} \\
&= \frac{\left(13\ \frac{\text{injury accidents}}{\text{yr}}\right)(10 \times 10^6)}{\left(6250\ \frac{\text{veh}}{\text{day}}\right)\left(365\ \frac{\text{day}}{\text{year}}\right)} \\
&= 57\ \text{accidents}/10^7\ \text{veh}
\end{aligned}
$$

The answer is (B).

189. Based on the percent of commuters, P_1, the number of commuting vehicles expected to use the parking garage is

$$
\begin{aligned}
V_1 &= P_1 V = (0.75)(400\ \text{veh}) \\
&= 300\ \text{veh}
\end{aligned}
$$

Based on the percent of shoppers, P_2, the number of shopping vehicles expected to use the parking garage is

$$
\begin{aligned}
V_2 &= P_2 V = (0.25)(400\ \text{veh}) \\
&= 100\ \text{veh}
\end{aligned}
$$

The total demand for the garage, D, is based on the number of vehicles and the time they spend parked. t_1 represents the average parking duration of the commuters, and t_2 is the average parking duration of the shoppers.

$$
\begin{aligned}
D &= V_1 t_1 + V_2 t_2 \\
&= (300\ \text{veh})(8\ \text{hr}) + (100\ \text{veh})(3\ \text{hr}) \\
&= 2700\ \text{veh-hr}
\end{aligned}
$$

Because each vehicle is expected to occupy one space, the demand can be expressed as 2700 spaces-hr. The total time the parking garage is available for parking, T, is from 6 a.m. to 5 p.m. (11 hr). f is the parking efficiency. N is the number of spaces required to meet the demand. The demand is equal to the product of the efficiency, the available time, and the number of spaces required to meet demand. The required supply, S, should be equal to or greater than the demand.

$$S \geq D = fTN$$

$$2700 \text{ spaces-hr} = (0.85)(11 \text{ hr})N$$

$$N = 289 \text{ spaces} \quad (290 \text{ spaces})$$

The answer is (C).

190. The width of available space for parking on each side of the road, W_P, is calculated from the width of the road, W, and the width of each lane, W_L.

$$W_P = \frac{W - 2W_L}{2}$$
$$= \frac{40 \text{ ft} - (2)(11 \text{ ft})}{2}$$
$$= 9 \text{ ft}$$

This parking width (9 ft) can only accommodate parallel parking, which is the parking configuration with the least traffic interference. The length of a parking space in a parallel parking configuration is based on a standard of 22 ft.

The total number of parallel parking spaces on both sides of the road is

$$N = \frac{(\text{no. of sides})(\text{road length})}{\text{stall length}}$$
$$= \frac{(2)(1.2 \text{ mi})\left(5280 \, \dfrac{\text{ft}}{\text{mi}}\right)}{22 \, \dfrac{\text{ft}}{\text{space}}}$$
$$= 576 \text{ spaces} \quad (580 \text{ spaces})$$

The answer is (B).

191. Follow the procedure given in *HCM* Exh. 12-6.

The weaving section is a one-sided ramp weave according to *HCM* Exh. 12-3(a), with a right-side single-lane on-ramp closely followed by a one-lane, right-side off-ramp, connected by a continuous freeway auxiliary lane.

Volume adjustments to correct for heavy vehicles, driver population, and PHF are not necessary as the volumes are passenger car equivalents, which already correct for these factors. From *HCM* Eq. 12.2, the minimum number of lane changes required is

$$LC_{RF} = 1$$

$$LC_{FR} = 1$$

$$LC_{min} = (LC_{RF})v_{RF} + (LC_{FR})v_{FR}$$
$$= (1)(300 \text{ pcph}) + (1)(200 \text{ pcph})$$
$$= 500 \text{ pcph}$$

The volume ratio is

$$VR = \frac{v_W}{v} = \frac{500 \text{ pcph}}{2500 \text{ pcph} + 2000 \text{ pcph} + 450 \text{ pcph}}$$
$$= 0.101$$

Find the maximum weaving segment. There are two weaving lanes, so $N_{WL} = 2$. From *HCM* Eq. 12-4, the maximum weaving segment length is

$$L_{max} = 5728(1 + VR)^{1.6} - 1566N_{WL}$$
$$= (5728)(1 + 0.101)^{1.6} - (1566)(2)$$
$$= 3549 \text{ ft}$$

The maximum weaving segment is consistent with *HCM* Exh. 12-9. The maximum weaving segment exceeds the available short length of weaving, so further analysis is needed.

Check the capacity of the weaving segment using *HCM* Eq. 12-5.

$$c_{IWL} = c_{IFL} - 438.2(1 + VR)^{1.6} + 0.0765L_S + 119.8N_{WL}$$
$$= 1023 \text{ pcphpl} - (438.2)(1 + 0.101)^{1.6}$$
$$\qquad + (0.0765)(1500 \text{ ft}) + (119.8)(2)$$
$$= 866 \text{ pcphpl}$$

From *HCM* Eq. 12-6, the total capacity is

$$c_W = c_{IWL}Nf_{HV}f_p = (866 \text{ pcphpl})(2)(1.0)(1.0)$$
$$= 1732 \text{ pcph}$$

Find the weaving segment capacity under equivalent ideal conditions using *HCM* Eq. 12-7. The capacity of the weaving segment is the smaller of c_{IWL} and c_{IW}.

$$c_{IW} = \frac{2400 \text{ pcph}}{VR} = \frac{2400 \text{ pcph}}{0.101} = 23{,}760 \text{ pcph}$$

The controlling capacity of the weaving segment is 1732 pcph.

Determine the total lane-changing rate using *HCM* Eq. 12-10 through Eq. 12-16. The equivalent lane-changing rate for vehicles within the weaving segment is

$$LC_W = LC_{min} + 0.39\left((L_S - 300)^{0.5}N^2(1 + ID)^{0.8}\right)$$

$$= 500 + (0.39)\left(\begin{array}{c}(1500 \text{ ft} - 300)^{0.5}\\ \times (2)^2(1 + 0.25)^{0.8}\end{array}\right)$$

$$= 565 \text{ pcph}$$

The nonweaving vehicle index is

$$I_{NW} = \frac{L_S(\text{ID})v_{NW}}{10{,}000} = \frac{(1500 \text{ ft})(0.25)(4450 \text{ pcph})}{10{,}000}$$

$$= 167 \quad [\le 1300, \text{ so Eq. 12-12 applies}]$$

The rate at which nonweaving vehicles in the segment change lanes is

$$LC_{NW} = 0.206v_{NW} + 0.542L_S - 192.6N$$

$$= (0.206)(4450 \text{ pcph}) + (0.542)(1500 \text{ ft})$$

$$\quad - (192.6)(2)$$

$$= 1344.5 \text{ pcph}$$

Determine the total lane changing rate. From *HCM* Eq. 12-16,

$$LC_{all} = LC_W + LC_{NW} = 565 \text{ pcph} + 1344.5 \text{ pcph}$$

$$= 1909.5 \text{ pcph}$$

Determine the average speed of weaving vehicles using *HCM* Eq. 12-17 and Eq. 12-19.

$$W = 0.226\left(\frac{LC_{all}}{L_S}\right)^{0.789} = (0.226)\left(\frac{1909.5 \text{ pcph}}{1500 \text{ pcph}}\right)^{0.789}$$

$$= 0.273$$

$$S_W = S_{min} + \frac{S_{max} - S_{min}}{1 + W}$$

$$= 15 \frac{\text{mi}}{\text{hr}} + \frac{60 \frac{\text{mi}}{\text{hr}} - 15 \frac{\text{mi}}{\text{hr}}}{1 + 0.273}$$

$$= 50 \text{ mph}$$

The answer is (C).

---

192. The total lost time for the whole intersection, L, is the sum of lost times for all the phases. From *HCM* Eq. 18-18,

$$L = \sum t_l = 4 \text{ sec} + 4 \text{ sec} + 2 \text{ sec} + 4 \text{ sec} = 14 \text{ sec}$$

Using *HCM* Eq. 18-17, the cycle length of the intersection traffic signal, C, can be determined given the desired critical ratio of flow to capacity for the overall intersection, X_c.

$$X = \left(\frac{C}{C - L}\right)\sum_{i \in ci} y_{c,i}$$

$$0.90 = (0.35 + 0.10 + 0.23 + 0.08)\left(\frac{C}{C - 14 \text{ sec}}\right)$$

$$C = 90 \text{ sec}$$

The answer is (C).

---

193. The problem statement provides information about the intersection: the maximum allowable speed, v; the width of intersection, W; the average length of vehicle, L; the perception-reaction time, t_{PR}; and the deceleration rate, a.

Using the ITE formula, the minimum yellow interval at this intersection is

$$t_{yellow,min} = t_{PR} + \frac{W + L}{v} + \frac{v}{2(a + Gg)}$$

$$= 2 \text{ sec} + \frac{(48 \text{ ft} + 20 \text{ ft})\left(3600 \frac{\text{sec}}{\text{hr}}\right)}{\left(45 \frac{\text{mi}}{\text{hr}}\right)\left(5280 \frac{\text{ft}}{\text{mi}}\right)}$$

$$+ \frac{\left(45 \frac{\text{mi}}{\text{hr}}\right)\left(5280 \frac{\text{ft}}{\text{mi}}\right)}{(2)\left(11.2 \frac{\text{ft}}{\text{sec}^2}\right)\left(3600 \frac{\text{sec}}{\text{hr}}\right)}$$

$$= 5.98 \text{ sec} \quad (6.0 \text{ sec})$$

The second term of $(W + L)/v$ is known as the "clearance time" since it represents the time required for a vehicle traveling at velocity, v, to go through the intersection. The clearance time was routinely included in the all-red interval up until approximately 1985. Cars traveling through the intersection under a red light were rare. Current practice, which results in more red light violations, is to exclude the clearance time from the all-red interval. In this solution, the clearance interval is specifically included in the all-red interval according to the problem statement.

v/2 is the average velocity traveled during deceleration.

The answer is (C).

194. Compound curves consist of two or more curves turning in the same direction while reverse curves turn in opposite directions. Therefore, the statement in option C is not true. All the other statements are true.

The answer is (C).

195. Rapid-curing asphalt is formed by cutting asphalt cement with a petroleum distillate, such as gasoline, that will easily evaporate.

The answer is (C).

196. A soil classified as A-7-6 (20) is usually rated "poor" as a subgrade and is considered unsuitable as a subbase material. Therefore, the statement in option D is not true. All other statements are true.

The answer is (D).

197. From the illustrations, for maximum unit weight, the asphalt content AC_{UW} is 5.4%. For maximum Marshall stability, the asphalt content AC_{ST} is 4.9%. For 4% total air voids (VTM), the asphalt content AC_{AI} is 4.5%. The optimum asphalt content AC_{OP} is

$$AC_{OP} = \frac{AC_{UW} + AC_{ST} + AC_{AI}}{3}$$

$$= \frac{5.4\% + 4.9\% + 4.5\%}{3}$$

$$= 4.93\% \quad (5.0\%)$$

The answer is (C).

198. There are several procedures for determining bulk specific gravity. For mixtures that are "open" (i.e., having significant porosity and permeability), or when highest accuracy is needed, the bulk specific gravity can be calculated from AASHTO T166, Method A, as

$$G_{mb} = \frac{m_{dry}}{m_{SSD} - m_{submerged}}$$

For routine testing of specimens with dense, impermeable surfaces, the dry and SSD masses are nearly identical, and the following formula can be used with sufficient accuracy for most purposes.

$$G_{mb} = \frac{m_{dry}}{m_{dry} - m_{submerged}}$$

$$= \frac{1238.5 \text{ g}}{1238.5 \text{ g} - 698.3 \text{ g}}$$

$$= 2.29$$

(The latter formula was used in this solution. In any case, the SSD mass was not given, so it would not have been possible to use Method A.)

The aggregate percent by weight of total paving mixture, P_s, is $100\% - 6\% = 94\%$. The percent VMA is

$$VMA = 100\% - \frac{G_{mb}P_s}{G_{sb}}$$

$$= 100\% - \frac{(2.29)(94\%)}{2.64}$$

$$= 18.37\% \quad (18\%)$$

The answer is (C).

199. The problem statement gives the following information about the project location and the pavement design: latitude, lat; the seven day average high air temperature, T_{high}; the one day average low air temperature, T_{low}; the standard deviation of the high temperature, σ_{high}; the standard deviation of the low temperature, σ_{low}; the pavement surface depth, H; and the reliability.

The seven day average high air temperature for 98% ($z = 2.05$; one-tail) reliability is

$$T_{high\,98} = T_{high} + 2.055\sigma_{high}$$

$$= 31°C + (2.055)(1.5°C)$$

$$= 34°C$$

The design pavement temperature at a depth of 20 mm is

$$T_{20\,mm} = \left(\begin{array}{c} T_{high\,98} - 0.00618(lat)^2 \\ + 0.2289(lat) + 42.2 \end{array} \right)(0.9545) - 17.78$$

$$= \left(\begin{array}{c} 34°C - (0.00618)(40.9°)^2 \\ + (0.2289)(40.9°) \\ + 42.2 \end{array} \right)(0.9545) - 17.78$$

$$= 54°C$$

The low pavement design temperature is

$$T_{pave} = 1.56 + 0.72\,T_{low} - 0.004(lat)^2$$

$$+ 6.26\log_{10}(H + 25)$$

$$- z\sqrt{4.4 + 0.52\sigma_{low}^2}$$

$$= 1.56 + (0.72)(-25°C) - (0.004)(40.9°)^2$$

$$+ 6.26\log_{10}(190 \text{ mm} + 25)$$

$$- 2.055\sqrt{4.4 + (0.52)(1.5°)^2}$$

$$= -13.4°C \quad (-13°C)$$

From a table that contains Superpave performance-graded asphalt binder specifications, use the computed values of $T_{20\,mm}$ and T_{pave} to select an appropriate performance grade asphalt binder for this project. The asphalt binder PG 58-16 is the most appropriate for 98% reliability.

The answer is (C).

200. The problem statement gives information about the constructed pavement, including its thickness, D; the layer coefficient, a; and the drainage coefficient, m. The materials used in the different layers—AC surface course, untreated granular base, and untreated gravel subbase—are represented by the subscripts 1, 2, and 3, respectively.

material	thickness, D (in)	layer coefficient, a	drainage coefficient, m
AC surface course, D_1	6	0.400	–
untreated granular base, D_2	8	0.115	0.50
untreated gravel subbase, D_3	10	0.090	0.50

The SN of the constructed pavement is

$$
\begin{aligned}
SN &= D_1 a_1 + D_2 a_2 m_2 + D_3 a_3 m_3 \\
&= (6\text{ in})(0.400) + (8\text{ in})(0.115)(0.50) \\
&\quad + (10\text{ in})(0.09)(0.50) \\
&= 3.31 \quad (3.3)
\end{aligned}
$$

The answer is (B).

Solutions
Afternoon Session
Water Resources and Environmental

201. The basin volume based on detention time is

$$V = Qt$$

$$= \frac{\left(2 \frac{\text{MG}}{\text{day}}\right)\left(10^6 \frac{\text{gal}}{\text{MG}}\right)(90 \text{ min})}{\left(7.48 \frac{\text{gal}}{\text{ft}^3}\right)\left(24 \frac{\text{hr}}{\text{day}}\right)\left(60 \frac{\text{min}}{\text{hr}}\right)}$$

$$= 16{,}711 \text{ ft}^3$$

The area based on detention time is

$$A = \frac{V}{h} = \frac{16{,}711 \text{ ft}^3}{8 \text{ ft}}$$

$$= 2089 \text{ ft}^2$$

The diameter based on detention time is

$$D = \sqrt{\frac{4A}{\pi}} = \sqrt{\frac{(4)(2089 \text{ ft}^2)}{\pi}}$$

$$= 51.6 \text{ ft}$$

The area based on overflow rate is

$$A = \frac{Q}{v^*} = \frac{(2 \text{ MGD})\left(10^6 \frac{\text{gal}}{\text{MG}}\right)}{600 \frac{\text{gal}}{\text{day-ft}^2}}$$

$$= 3333 \text{ ft}^2$$

The diameter based on overflow rate is

$$D = \sqrt{\frac{4A}{\pi}} = \sqrt{\frac{(4)(3333 \text{ ft}^2)}{\pi}}$$

$$= 65.1 \text{ ft} \quad [\text{controls}]$$

Check the weir loading for a 65 ft diameter. The weir length is

$$L = \pi D = \pi(65 \text{ ft})$$

$$= 204.2 \text{ ft}$$

The weir loading is

$$q_{\text{weir}} = \frac{Q}{L} = \frac{(2 \text{ MGD})\left(10^6 \frac{\text{gal}}{\text{MG}}\right)}{204.2 \text{ ft}}$$

$$= 9794 \text{ gal/day-ft}$$

The weir loading is acceptable. The diameter of the basin is 65 ft.

The answer is (C).

202. The mass of available alum is

$$m_{\text{alum}} = C_{\text{alum}} P$$

C_{alum} is the concentration of alum, and P is the fraction of available alum.

$$m_{\text{alum}} = \left(1.400 \frac{\text{kg}}{\text{L}}\right)(0.49)$$

$$= 0.686 \text{ kg/L}$$

The atomic weight of aluminum is 26.98 g/mol. The aluminum available is

$$C_{\text{Al}} = m_{\text{alum}}(\text{WR})$$

WR is the ratio of mass of aluminum to mass of available alum.

$$C_{\text{Al}} = \left(0.686 \frac{\text{kg alum}}{\text{L solution}}\right)\left(2 \frac{\text{mol Al}}{\text{mol alum}}\right)$$

$$\times \left(26.98 \frac{\text{g Al}}{\text{mol Al}}\right)\left(\frac{1 \text{ mol alum}}{666.7 \text{ g alum}}\right)$$

$$\times \left(1000 \frac{\text{g}}{\text{kg}}\right)$$

$$= 55.5 \text{ g Al/L solution}$$

The reaction for phosphate precipitation with aluminum is

$$\text{Al}^{+3} + \text{H}_n\text{PO}_4^{3-n} \rightleftharpoons \text{AlPO}_4 + n\text{H}^+$$

One mole of aluminum is required to precipitate 1 mol of phosphorus.

The atomic weight of phosphorus is 30.97 g/mol. The theoretical unit alum dose required is

$$D_{\text{alum}} = \frac{m_{\text{Al}}}{m_{\text{P}}}$$
$$= \left(\frac{1 \text{ mol Al}}{1 \text{ mol P}}\right)\left(\frac{1 \text{ mol P}}{30.97 \text{ g P}}\right)\left(26.98 \frac{\text{g Al}}{\text{mol Al}}\right)$$
$$= 0.871 \text{ g Al/g P}$$

R is the ratio of actual to theoretical dose required. The unit volume of alum required is

$$V_{\text{unit}} = \frac{D_{\text{alum}} R}{C_{\text{Al}}}$$
$$= \frac{\left(0.871 \frac{\text{g Al}}{\text{g P}}\right)\left(1.50 \frac{\text{actual dose}}{\text{theoretical dose}}\right)}{55.5 \frac{\text{g Al}}{\text{L alum solution}}}$$
$$= 0.0235 \text{ L alum solution/g P}$$

The volume of alum solution required is

$$V_{\text{total}} = QCV_{\text{unit}}$$
$$= \frac{\left(400 \frac{\text{L}}{\text{s}}\right)\left(86\,400 \frac{\text{s}}{\text{d}}\right)\left(10 \frac{\text{mg P}}{\text{L}}\right)}{1000 \frac{\text{mg}}{\text{g}}}$$
$$= 8122 \text{ L/d alum solution} \quad (8000 \text{ L/d})$$

The answer is (B).

203. The overall mass of TSS removed is

$$m_{\text{TSS}} = (C_i - C_e)Q$$

C_i is the influent TSS concentration in mg/L, C_e is the effluent TSS concentration in mg/L, and Q is the flow in L/s.

The influent TSS is

$$\left(\frac{100\% - 75\%}{100\%}\right)\left(220 \frac{\text{mg}}{\text{L}}\right) = 55 \text{ mg/L}$$

$$m_{\text{TSS}} = \frac{\left(55 \frac{\text{mg}}{\text{L}} - 20 \frac{\text{mg}}{\text{L}}\right)\left(400 \frac{\text{L}}{\text{s}}\right)\left(86\,400 \frac{\text{s}}{\text{d}}\right)}{10^6 \frac{\text{mg}}{\text{kg}}}$$
$$= 1210 \text{ kg/d}$$

The BOD entering the aeration process is

$$C_a = C_i F$$

F is the fraction of BOD entering aeration process.

$$C_a = \left(280 \frac{\text{mg}}{\text{L}}\right)\left(\frac{100\% - 30\%}{100\%}\right)$$
$$= 196 \text{ mg/L}$$

The mass of biological solids produced is

$$m_{\text{bio}} = Y(\Delta\text{BOD})Q$$

Y is the cell yield in kg/kg.

$$m_{\text{bio}} = \frac{\left(\frac{60 \text{ kg suspended solids removed}}{100 \text{ kg BOD removed}}\right) \times \left(196 \frac{\text{mg}}{\text{L}} \text{ BOD} - 20 \frac{\text{mg}}{\text{L}} \text{ BOD}\right) \times \left(400 \frac{\text{L}}{\text{s}}\right)\left(86\,400 \frac{\text{s}}{\text{d}}\right)}{10^6 \frac{\text{mg}}{\text{kg}}}$$
$$= 3650 \text{ kg suspended biological solids}$$

The total dry sludge mass produced in the activated sludge process is

$$m_{\text{total}} = m_{\text{TSS}} + m_{\text{bio}}$$
$$= 1210 \frac{\text{kg}}{\text{d}} + 3650 \frac{\text{kg}}{\text{d}}$$
$$= 4860 \text{ kg/d} \quad (5000 \text{ kg/d})$$

The answer is (A).

204. The hydraulic detention time is

$$t = \frac{Lwnd}{Q}$$

L is the basin length, w is the basin width, n is the fraction of the cross-sectional area not occupied by plants, d is the basin depth, and Q is the average flow rate in ft^3/day.

$$t = \frac{(200 \text{ ft})(2500 \text{ ft})(0.75)(2 \text{ ft})}{40{,}000 \frac{\text{ft}^3}{\text{day}}}$$
$$= 18.75 \text{ days}$$

The effluent BOD is

$$C_e = C_o A e^{-0.7 K_T A_v^{1.75} t}$$

C_o is the influent BOD_5, A is the coefficient that represents the fraction of BOD_5 not removed by settling at the head of the system, K_T is the temperature-dependent, first-order rate constant in d^{-1}, and A_v is the specific surface area for microbiological activity in ft^2/ft^3.

$$K_T = K_{20°C}(1.047)^{T-20°C}$$
$$= (0.006 \text{ d}^{-1})(1.047)^{10°C-20°C}$$
$$= 0.0038 \text{ d}^{-1}$$

$$C_e = \left(200 \ \frac{mg}{L}\right)(0.52)e^{\left(\begin{array}{c}(-0.7)(0.0038 \text{ d}^{-1}) \\ \times (4.8 \text{ ft}^2/\text{ft}^3)^{1.75}(18.75 \text{ d})\end{array}\right)}$$
$$= 47.9 \text{ mg/L} \quad (48 \text{ mg/L})$$

The answer is (A).

205. The stream velocity is

$$v = \frac{Q}{A} = \frac{400 \ \dfrac{L}{s}}{(1 \text{ m})(10 \text{ m})\left(1000 \ \dfrac{L}{m^3}\right)}$$
$$= 0.04 \text{ m/s}$$

The travel time is

$$t = \frac{L}{v}$$
$$= \frac{(10 \text{ km})\left(1000 \ \dfrac{m}{km}\right)}{\left(0.04 \ \dfrac{m}{s}\right)\left(60 \ \dfrac{s}{min}\right)\left(60 \ \dfrac{min}{h}\right)}$$
$$= 69.4 \text{ h}$$

The decimal fraction of bacteria remaining is

$$\frac{N}{N_o} = (1 + nkt)^{-1/n}$$

N is the modal number of bacteria remaining, N_o is the modal number of initial bacteria, n is the coefficient of nonuniformity or retardation for the specific watercourse, and k is the intial die-away for the specific bacteria population in the receiving water.

$$\frac{N}{N_o} = \left(1 + (6.15)\left(\frac{1500}{h}\right)(69.4 \text{ h})\right)^{-1/6.15}$$
$$= 0.11$$

The percent removal is

$$P = \left(1 - \frac{N}{N_o}\right) \times 100\%$$
$$= (1 - 0.11) \times 100\%$$
$$= 89\% \quad (90\%)$$

The answer is (D).

206. The overall denitrification rate is

$$U'_{dn} = U_{dn}1.09^{T-20°C}(1 - DO)$$

U_{dn} is the specific denitrification rate in (kg NO_3-N)/(kg·d MLVSS), T is the wastewater temperature in °C, and DO is the wastewater dissolved oxygen in mg/L.

$$U'_{dn} = \left(0.09 \ \frac{\text{kg NO}_3\text{-N}}{\text{kg·d MLVSS}}\right)(1.09^{10°C-20°C})$$
$$\times \left(1 \ \frac{mg}{L} - 0.2 \ \frac{mg}{L}\right)$$
$$= 0.030 \text{ kg NO}_3\text{-N/kg·d MLVSS}$$

The detention time is

$$\theta = \frac{S_o - S}{U'_{dn}X}$$

S_o is the influent substrate concentration in mg/L, S is the effluent substrate concentration in mg/L, and X is the MLVSS in mg/L.

$$\theta = \frac{\left(26 \ \dfrac{mg}{L} - 4 \ \dfrac{mg}{L}\right)\left(24 \ \dfrac{h}{d}\right)}{\left(0.030 \ \dfrac{\text{kg NO}_3\text{-N}}{\text{kg·d MLVSS}}\right)\left(2500 \ \dfrac{mg}{L}\right)}$$
$$= 7.04 \text{ h} \quad (7 \text{ h})$$

The answer is (C).

207. The BOD_5 of the stream after complete mixing with the effluent is

$$C_f = \frac{C_1Q_1 + C_2Q_2}{Q_1 + Q_2}$$

C_f is the final mixed concentration, C_1 is the effluent concentration, C_2 is the stream concentration before mixing, Q_1 is the effluent flow, and Q_2 is the stream flow before mixing.

$$BOD_{5,20°C} = \frac{\left(60 \ \dfrac{mg}{L}\right)\left(0.5 \ \dfrac{m^3}{s}\right) + \left(4.5 \ \dfrac{mg}{L}\right)\left(6 \ \dfrac{m^3}{s}\right)}{0.5 \ \dfrac{m^3}{s} + 6 \ \dfrac{m^3}{s}}$$
$$= 8.77 \text{ mg/L}$$

Use the same equation twice more to find the DO and the temperature of the stream after complete mixing with the effluent. For the variables C_f, C_1, and C_2, substitute the appropriate values of DO and temperature, respectively. The DO of the stream after complete mixing with the effluent is

$$DO_{mix} = \frac{\left(2 \; \frac{mg}{L}\right)\left(0.5 \; \frac{m^3}{s}\right) + \left(8.5 \; \frac{mg}{L}\right)\left(6 \; \frac{m^3}{s}\right)}{0.5 \; \frac{m^3}{s} + 6 \; \frac{m^3}{s}}$$

$$= 8.0 \; mg/L$$

The temperature of the stream after complete mixing is

$$T = \frac{(25°C)\left(0.5 \; \frac{m^3}{s}\right) + (15°C)\left(6 \; \frac{m^3}{s}\right)}{0.5 \; \frac{m^3}{s} + 6 \; \frac{m^3}{s}}$$

$$= 15.77°C$$

The deoxygenation rate constant after complete mixing is

$$K_{d,T} = K_{d,20°C}(1.047)^{T-20°C}$$

$$= (0.15 \; d^{-1})(1.047)^{15.77°C-20°C}$$

$$= 0.124 \; d^{-1}$$

The reaeration rate constant after complete mixing is

$$K_{r,T} = K_{r,20°C}(1.024)^{T-20°C}$$

$$= (0.25 \; d^{-1})(1.024)^{15.77°C-20°C}$$

$$= 0.226 \; d^{-1}$$

The ultimate BOD of the stream after complete mixing at 20°C is

$$BOD_{u,20°C} = \frac{BOD_{5,20°C}}{1 - 10^{-K_d t}}$$

$$= \frac{8.77 \; \frac{mg}{L}}{1 - 10^{-(0.15 \; d^{-1})(5 \; d)}}$$

$$= 10.67 \; mg/L$$

The ultimate BOD of the stream after complete mixing at 15.77°C is

$$BOD_{u,T} = BOD_{u,20°C}(0.02T + 0.6)$$

$$= \left(10.67 \; \frac{mg}{L}\right)\left((0.02)(15.77°C) + 0.6\right)$$

$$= 9.77 \; mg/L$$

The oxygen deficit after complete mixing is

$$D_0 = DO_{sat} - DO_{mix}$$

DO_{sat} is the dissolved oxygen saturation, and DO_{mix} is the dissolved oxygen after complete mixing.

$$D_0 = 9.95 \; \frac{mg}{L} - 8.0 \; \frac{mg}{L}$$

$$= 1.95 \; mg/L$$

The time to the critical point is

$$t_c = \left(\frac{1}{K_r - K_d}\right)$$

$$\times \log_{10}\left(\left(\frac{K_d BOD_u - K_r D_0 + k_d D_0}{K_d BOD_u}\right)\left(\frac{K_r}{K_d}\right)\right)$$

$$= \left(\frac{1}{0.226 \; d^{-1} - 0.124 \; d^{-1}}\right)$$

$$\times \log_{10}\left(\frac{\begin{pmatrix}(0.124 \; d^{-1})\left(9.77 \; \frac{mg}{L}\right) \\ - (0.226 \; d^{-1})\left(1.95 \; \frac{mg}{L}\right) \\ + (0.124 \; d^{-1})\left(1.95 \; \frac{mg}{L}\right)\end{pmatrix}}{(0.124 \; d^{-1})\left(9.77 \; \frac{mg}{L}\right)}\right.$$

$$\left.\times \left(\frac{0.226 \; d^{-1}}{0.124 \; d^{-1}}\right)\right)$$

$$= 1.792 \; d$$

The critical oxygen deficit is

$$D_c = \left(\frac{K_d BOD_u}{K_r}\right)10^{-K_d t_c}$$

$$= \left(\frac{(0.124 \; d^{-1})\left(9.77 \; \frac{mg}{L}\right)}{0.226 \; d^{-1}}\right)$$

$$\times \left(10^{-(0.124 \; d^{-1})(1.79 \; d)}\right)$$

$$= 3.22 \; mg/L$$

The DO at the critical point is

$$DO = DO_{sat} - D_c$$

$$= 9.95 \; \frac{mg}{L} - 3.22 \; \frac{mg}{L}$$

$$= 6.73 \; mg/L \quad (6.8 \; mg/L)$$

The answer is (C).

208. The detention time is

$$t = \frac{V}{Q} = \frac{(1440 \text{ m}^3)\left(1000 \frac{\text{L}}{\text{m}^3}\right)}{\left(400 \frac{\text{L}}{\text{s}}\right)\left(60 \frac{\text{s}}{\text{min}}\right)}$$

$$= 60 \text{ min}$$

The required CT value, $C_t t$, can be found with the following equation.

$$\frac{N_t}{N_o} = (1 + 0.23 C_t t)^{-3}$$

N_t is the number of coliform organisms after treatment at time t, N_o is the number of coliform organisms before treatment, and C_t is the total amperometric chlorine residual at time t in mg/L.

$$\frac{\dfrac{200}{100 \text{ mL}}}{\dfrac{2 \times 10^8}{100 \text{ mL}}} = (1 + 0.23 C_t t)^{-3}$$

$$C_t t = 430.4 \text{ mg/L·min}$$

For a time of 60 min, C_t is

$$C_t = \frac{430.4 \frac{\text{mg}}{\text{L·min}}}{60 \text{ min}}$$

$$= 7.17 \text{ mg/L} \quad (7.2 \text{ mg/L})$$

The answer is (D).

209. When more than three dilutions are employed in a decimal series of dilutions, the results from only three of these are used in computing the MPN. The three dilutions selected are the highest dilution (i.e., the lowest sample portion) giving positive results in all five portions tested (no lower dilution with any negative results), and the two next succeedingly higher dilutions. The dilution corresponding to the middle dilution is used to calculate the MPN from the MPN index. The three dilutions used are given in the following table.

serial dilution	sample portion (mL)	number of positive reactions
2	0.01	5
3	0.001	2
4	0.0001	1

The middle dilution is 0.001 mL. Using the MPN tables, the MPN index is 70 MPN/100 mL.

The MPN is

$$\text{MPN} = \frac{I}{D}$$

I is the MPN index in MPN/100 mL. D is the middle dilution corresponding to 1 mL of a series of 10 mL, 1 mL, and 0.1 mL.

$$\text{MPN} = \frac{\dfrac{70 \text{ MPN}}{100 \text{ mL}}}{0.001 \dfrac{\text{mL}}{\text{mL}}}$$

$$= 70\,000/100 \text{ mL}$$

The answer is (D).

210. The decimal fraction of the wastewater sample used for dilution no. 1 is

$$P = \frac{S}{T}$$

S is the sample volume. T is the total volume of the bottle.

$$P = \frac{5 \text{ mL}}{300 \text{ mL}}$$

$$= 0.0167$$

The decimal fraction of the wastewater sample used for dilution no. 2 is

$$P = \frac{10 \text{ mL}}{300 \text{ mL}}$$

$$= 0.0333$$

The decimal fraction of the wastewater sample used for dilution no. 3 is

$$P = \frac{15 \text{ mL}}{300 \text{ mL}}$$

$$= 0.050$$

The 5 d BOD is

$$\text{BOD}_5 = \frac{\text{DO}_i - \text{DO}_f}{P}$$

DO_i is the initial DO in mg/L, and DO_f is the final DO after 5 d in mg/L. The 5 d BOD of dilution no. 1 is

$$\text{BOD}_5 = \frac{8.0 \frac{\text{mg}}{\text{L}} - 6.2 \frac{\text{mg}}{\text{L}}}{0.0167}$$

$$= 107.8 \text{ mg/L}$$

The 5 d BOD of dilution no. 2 is

$$\text{BOD}_5 = \frac{8.2 \ \dfrac{\text{mg}}{\text{L}} - 5.2 \ \dfrac{\text{mg}}{\text{L}}}{0.0333}$$
$$= 90.0 \text{ mg/L}$$

The 5 d BOD of dilution no. 3 is

$$\text{BOD}_5 = \frac{8.4 \ \dfrac{\text{mg}}{\text{L}} - 3.5 \ \dfrac{\text{mg}}{\text{L}}}{0.050}$$
$$= 98.0 \text{ mg/L}$$

The average 5 d BOD of the three dilutions is

$$\text{BOD}_{\text{avg}} = \frac{\sum \text{BOD}}{n}$$

BOD_n represents individual BOD results. n is the number of samples.

$$\text{BOD}_{\text{avg}} = \frac{107.8 \ \dfrac{\text{mg}}{\text{L}} + 90.0 \ \dfrac{\text{mg}}{\text{L}} + 98.0 \ \dfrac{\text{mg}}{\text{L}}}{3}$$
$$= 98.6 \text{ mg/L}$$

The ultimate BOD is

$$\text{BOD}_u = \frac{\text{BOD}_5}{1 - e^{-k_1 t}}$$

k_1 is the rate constant to base e, and t is the test duration in d.

$$\text{BOD}_u = \frac{98.6 \ \dfrac{\text{mg}}{\text{L}}}{1 - e^{-(0.25 \text{ d}^{-1})(5 \text{ d})}}$$
$$= 138 \text{ mg/L} \quad (140 \text{ mg/L})$$

The answer is (C).

211. The direct haul cost per unit volume and unit time is

$$C'_{\text{dh}} = \frac{C_{\text{dh}}}{L_{\text{dh}}}$$

C_{dh} is the cost of direct haul per minute, and L_{dh} is the loading capacity of direct haul.

$$C'_{\text{dh}} = \frac{\left(40 \ \dfrac{\$}{\text{hr}}\right)\left(\dfrac{1 \text{ direct haul vehicle}}{15 \text{ yd}^3}\right)}{60 \ \dfrac{\text{min}}{\text{hr}}}$$
$$= \$0.044/\text{yd}^3\text{-min}$$

The semi-trailer haul cost per unit volume and unit time is

$$C'_{\text{st}} = \frac{C_{\text{st}}}{L_{\text{st}}}$$

C_{st} is the cost of semi-trailer haul per minute, and L_{st} is the loading capacity of semi-trailer.

$$C'_{\text{st}} = \frac{\left(50 \ \dfrac{\$}{\text{hr}}\right)\left(\dfrac{1 \text{ semi-trailer vehicle}}{100 \text{ yd}^3}\right)}{60 \ \dfrac{\text{min}}{\text{hr}}}$$
$$= \$0.0083/\text{yd}^3\text{-min}$$

Plot the round trip time versus the cost per cubic yard for each system. C'_{dh} and C'_{st} are the line slopes. Start the semi-trailer plot at the transfer station operation cost.

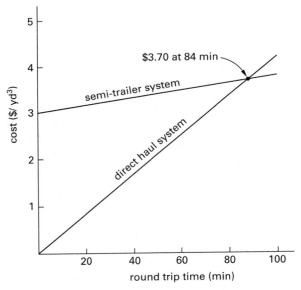

The direct haul cost is equivalent to the cost of the semi-trailer system at a round trip time of 84 min (85 min).

The answer is (A).

212. The frequency of exposure is

$$\text{FOE} = \frac{t_{\text{exposure}}}{(1 \text{ yr})\left(365 \, \frac{\text{d}}{\text{yr}}\right)}$$

t_{exposure} is the exposure time measured in days.

$$\text{FOE} = \frac{120 \text{ d}}{365 \text{ d}}$$
$$= 0.329$$

The absorbed dose, AB, is 1.0.

The drinking water equivalent level in mg/L is

$$\text{DWEL} = \frac{D_{\text{ref}}(\text{ABW})}{(\text{DWI})(\text{AB})(\text{FOE})}$$

D_{ref} is the oral reference dose for MTBE in mg/kg·d, ABW is the average body mass in kg, and DWI is the daily water intake in L.

$$\text{DWEL} = \frac{\left(0.005 \, \frac{\text{mg}}{\text{kg·d}}\right)(70 \text{ kg})}{\left(2 \, \frac{\text{L}}{\text{d}}\right)(1.0)(0.333)}$$
$$= 0.53 \text{ mg/L}$$

The answer is (C).

213. The thickness of the saturated aquifier at the radius of influence is

$$y_1 = y_{\text{inf}} - y_{\text{bottom}}$$

y_{inf} is the phreatic surface elevation at the radius of influence, and y_{bottom} is the elevation of the bottom of the aquifer.

$$y_1 = 100 \text{ m} - 55 \text{ m} = 45 \text{ m}$$

The thickness of the saturated aquifier at the well is

$$y_2 = y_{\text{inf}} - y_{\text{bottom}} - y_{\text{drawdown}}$$

y_{drawdown} is the drawdown in the well.

$$y_2 = 100 \text{ m} - 55 \text{ m} - 3.5 \text{ m} = 41.5 \text{ m}$$

The radius of the well is

$$r_2 = \frac{D_{\text{well}}}{2}$$

D_{well} is the diameter of the well.

$$r_2 = \left(\frac{450 \text{ mm}}{2}\right)\left(\frac{1 \text{ m}}{1000 \text{ mm}}\right)$$
$$= 0.225 \text{ m}$$

Using Dupuit's equation, the hydraulic conductivity is

$$K = \frac{Q \ln \dfrac{r_1}{r_2}}{\pi(y_1^2 - y_2^2)}$$

$$= \frac{\left(50 \, \frac{\text{L}}{\text{s}}\right)\ln\left(\dfrac{300 \text{ m}}{0.225 \text{ m}}\right)}{\pi\left((45 \text{ m})^2 - (41.5 \text{ m})^2\right)\left(1000 \, \frac{\text{L}}{\text{m}^3}\right)}$$

$$= 378.26 \times 10^{-6} \text{ m/s} \quad (400 \times 10^{-6} \text{ m/s})$$

The answer is (B).

214. Total hardness includes primarily calcium and magnesium ions, and to a lesser extent (because of normally lower concentrations) iron, manganese, strontium, and aluminum. The concentrations of hardness-causing ions are

$$\text{Ca}^{2+} = 90 \text{ mg/L}$$
$$\text{Mg}^{2+} = 38 \text{ mg/L}$$
$$\text{Fe}^{2+} = 0.10 \text{ mg/L}$$
$$E = \frac{M}{Z}$$

E is the equivalent weight, and M is the molecular weight. Z is the absolute value of the ion change: the number of H^+ or OH^- ions a species can react with or yield in an acid-base reaction, or the absolute value of the change in valence occcurring in an oxidation-reduction reaction.

The millequivalents are

$$\text{Ca}^{2+} = \frac{40 \text{ mg}}{2} = 20 \text{ mg/meq}$$
$$\text{Mg}^{2+} = \frac{24 \text{ mg}}{2} = 12 \text{ mg/meq}$$
$$\text{Fe}^{2+} = \frac{56 \text{ mg}}{2} = 28 \text{ mg/meq}$$

The hardness ions in meq/L are

$$H = \frac{C}{E}$$

H is the hardness in meq/L, C is the concentration in mg/L, and E is the millequivalents of hardness-causing ions.

$$Ca^{2+} = \frac{90 \ \frac{mg}{L}}{20 \ \frac{mg}{meq}} = 4.50 \ meq/L$$

$$Mg^{2+} = \frac{38 \ \frac{mg}{L}}{12 \ \frac{mg}{meq}} = 3.17 \ meq/L$$

$$Fe^{2+} = \frac{0.10 \ \frac{mg}{L}}{28 \ \frac{mg}{meq}} = 0.0036 \ meq/L$$

The millequivalent weight of $CaCO_3$ is

$$E_{CaCO_3} = \frac{40 \ mg + 12 \ mg + (3)(16 \ mg)}{2}$$
$$= 50 \ mg/meq$$

The total hardness as $CaCO_3$ is

$$\left(4.50 \ \frac{meq}{L} + 3.17 \ \frac{meq}{L} + 0.0036 \ \frac{meq}{L} \right)$$
$$\times \left(50 \ \frac{mg \ CaCO_3}{meq} \right)$$
$$= 384 \ mg/L \ as \ CaCO_3 \quad (380 \ mg/L \ as \ CaCO_3)$$

The answer is (A).

215. The soil permeability is

$$K = \left(350 \ \frac{m}{yr} \right) \left(\frac{1 \ yr}{365 \ d} \right) \left(\frac{1 \ d}{86 \ 400 \ s} \right)$$
$$= 1.1 \times 10^{-5} \ m/s$$

The total flow for each drain is given by the rational equation for radial flow. L is the length of the drain in meters, H is the height of the water table above the invert of drain at the midpoint of drain spacing in meters, h is the height of the water level above the drain invert in the pipe in meters, S is the spacing between drains in meters, and D is the diameter of the drains in meters.

$$Q = \frac{\pi K L (H - h)}{\ln \frac{S}{D}}$$
$$= \frac{\pi \left(1.1 \times 10^{-5} \ \frac{m}{s} \right) (1200 \ m)(3.5 \ m - 0.3 \ m)}{\ln \left(\frac{100 \ m}{0.3 \ m} \right)}$$
$$= 0.022 \ 84 \ m^3/s \quad (0.025 \ m^3/s)$$

The answer is (C).

216. The ratio of the design head to the test head is

$$\frac{H}{H_o} = \frac{16 \ ft}{10 \ ft} = 1.6$$

From the test results, the coefficient of discharge is

$$\frac{C}{C_o} = 1.09$$
$$C = 1.09 C_o = (1.09) \left(2.20 \ \frac{ft^{1/2}}{sec} \right)$$
$$= 2.40 \ ft^{1/2}/sec$$

The discharge for a broad-crested weir is

$$Q = C b H^{3/2}$$
$$= \left(2.40 \ \frac{ft^{1/2}}{sec} \right) (16 \ ft)(16 \ ft)^{3/2}$$
$$= 2457 \ ft^3/sec \quad (2500 \ ft^3/sec)$$

The answer is (B).

217. The total available energy head is equal to the velocity head at the discharge plus the total losses in the piping system. The losses include friction at the square entrance, h_e; friction in pipe section 1-2, h_{f1}; sudden contraction, h_c; friction in pipe section 2-3, h_{f2}; friction in the rotary valve, h_{valve}; and the loss at the free discharge, h_d. The minor losses are expressed using a loss coefficient, K, of the velocity head. The head loss for a square edge entrance ($K_e = 0.5$) is

$$h_e = K_e h_v = K_e \frac{v_1^2}{2 g_c}$$
$$= 0.5 \frac{v_1^2}{2 g_c}$$

The head loss for pipe section 1-2 is

$$h_{f_1} = f_1 \left(\frac{L}{D} \right) \left(\frac{v_1^2}{2g} \right) = f_1 \left(\frac{3940 \ ft}{20 \ in} \right) \left(\frac{v_1^2}{2g} \right) \left(12 \ \frac{in}{ft} \right)$$
$$= 2364 f_1 \frac{v_1^2}{2g}$$

The head loss for the sudden contraction is

$$K_c = \tfrac{1}{2} \left(1 - \left(\frac{D_s}{D_l} \right)^2 \right)$$
$$= \left(\tfrac{1}{2} \right) \left(1 - \left(\frac{12 \ in}{20 \ in} \right)^2 \right)$$
$$= 0.32$$
$$h_c = 0.32 \frac{v_2^2}{2g}$$

The head loss for pipe section 2-3 is

$$h_{f_2} = f_2 \left(\frac{L}{D}\right)\left(\frac{v_2^2}{2g}\right)$$

$$= f_2 \left(\frac{4920 \text{ ft}}{12 \text{ in}}\right)\left(\frac{v_2^2}{2g}\right)\left(12 \frac{\text{in}}{\text{ft}}\right)$$

$$= 4920 f_2 \frac{v_2^2}{2g}$$

The head loss for a fully open rotary valve is

$$h_{\text{valve}} = \frac{10 v_2^2}{2g}$$

The head loss for the free discharge is

$$h_d = \frac{v_2^2}{2g}$$

From the continuity equation,

$$v_1 = \frac{A_2 v_2}{A_1} = \frac{D_2^2 v_2}{D_1^2}$$

$$= \frac{(12 \text{ in})^2 v_2}{(20 \text{ in})^2}$$

$$= 0.36 v_2$$

To evaluate f_1 and f_2, the Reynolds numbers for points 1 and 2 are needed.

$$\text{Re}_1 = \frac{D_1 v_1}{\nu} = \frac{(20 \text{ in})v_1}{\left(1.410 \times 10^{-5} \frac{\text{ft}^2}{\text{sec}}\right)\left(12 \frac{\text{in}}{\text{ft}}\right)}$$

$$= 1.18 \times 10^5 v_1$$

$$\text{Re}_2 = \frac{D_2 v_2}{\nu} = \frac{(12 \text{ in})v_2}{\left(1.410 \times 10^{-5} \frac{\text{ft}^2}{\text{sec}}\right)\left(12 \frac{\text{in}}{\text{ft}}\right)}$$

$$= 7.09 \times 10^4 v_2$$

The relative roughness for each of the two pipes is

$$\frac{\epsilon_1}{D_1} = \left(\frac{0.004 \text{ ft}}{20 \text{ in}}\right)\left(12 \frac{\text{in}}{\text{ft}}\right)$$

$$= 0.0024$$

$$\frac{\epsilon_2}{D_2} = \left(\frac{0.004 \text{ ft}}{12 \text{ in}}\right)\left(12 \frac{\text{in}}{\text{ft}}\right)$$

$$= 0.004$$

Solve by iteration. For trial 1, assume f_1 is 0.022 and f_2 is 0.028. The head loss relationships are

$$\Delta E_k = \Delta E_p$$

$$h_e + h_{f1} + h_c + h_{f2} + h_{\text{valve}} + h_d + h_v = 295 \text{ ft}$$

$$(0.5 + 2364 f_1)\left(\frac{(0.36 v_2)^2}{(2)\left(32.2 \frac{\text{ft}}{\text{sec}^2}\right)}\right)$$

$$+ \left(\begin{array}{c} 0.32 + 4920 f_2 \\ + 10 + 1 + 1 \end{array}\right)$$

$$\times \left(\frac{v_2^2}{(2)\left(32.2 \frac{\text{ft}}{\text{sec}^2}\right)}\right) = 295 \text{ ft}$$

Solve for v_2 using $f_1 = 0.022$ and $f_2 = 0.028$.

$$v_2 = 11.01 \text{ ft/sec}$$

$$v_1 = 0.36 v_2 = (0.36)\left(11.01 \frac{\text{ft}}{\text{sec}}\right)$$

$$= 3.96 \text{ ft/sec}$$

The Reynolds numbers are

$$\text{Re}_1 = \frac{(20 \text{ in})\left(3.96 \frac{\text{ft}}{\text{sec}}\right)}{\left(1.410 \times 10^{-5} \frac{\text{ft}^2}{\text{sec}}\right)\left(12 \frac{\text{in}}{\text{ft}}\right)}$$

$$= 4.68 \times 10^5$$

$$\text{Re}_2 = \frac{(12 \text{ in})\left(11.01 \frac{\text{ft}}{\text{sec}}\right)}{\left(1.410 \times 10^{-5} \frac{\text{ft}^2}{\text{sec}}\right)\left(12 \frac{\text{in}}{\text{ft}}\right)}$$

$$= 7.81 \times 10^5$$

From a Moody diagram, the friction factors are $f_1 = 0.025$ and $f_2 = 0.029$. These values are close enough to the assumed friction factors used to find the pipe velocities. The velocities in the pipe are

$$v_1 = 4 \text{ ft/sec}$$

$$v_2 = 11 \text{ ft/sec}$$

The discharge is

$$Q = A_2 v_2 = \frac{D_2^2}{4} v_2$$

$$= \frac{\pi \left(\frac{12 \text{ in}}{12 \frac{\text{in}}{\text{ft}}}\right)^2 \left(11 \frac{\text{ft}}{\text{sec}}\right)}{4}$$

$$= 8.6 \text{ ft}^3/\text{sec} \quad (9.0 \text{ ft}^3/\text{sec})$$

The answer is (B).

218. The slope of the stream is

$$S_0 = \frac{\Delta z}{L} = \frac{1.54 \text{ ft}}{2135.69 \text{ ft}}$$
$$= 0.00072 \text{ ft/ft}$$

The Manning roughness coefficient, n, for a natural channel with stones and weeds is 0.035. The maximum flow area is

$$A = dw = (6 \text{ ft})(4 \text{ ft})$$
$$= 24 \text{ ft}^2$$

The wetted perimeter is

$$P = 2d + w = (2)(6 \text{ ft}) + 4 \text{ ft}$$
$$= 16 \text{ ft}$$

The hydraulic radius is

$$R = \frac{A}{P} = \frac{24 \text{ ft}^2}{16 \text{ ft}}$$
$$= 1.5 \text{ ft}$$

The flow from the Chezy-Manning equation is

$$Q = \left(\frac{1.49}{n}\right) A R^{2/3} \sqrt{S_0}$$
$$= \left(\frac{1.49}{0.035}\right)(24 \text{ ft}^2)(1.5 \text{ ft})^{2/3}\sqrt{0.00072 \ \frac{\text{ft}}{\text{ft}}}$$
$$= 35.9 \text{ ft}^3/\text{sec} \quad (36 \text{ ft}^3/\text{sec})$$

The answer is (A).

219. Determine the system head curves for the two clear well elevations. Then, plot the pump curve against the system head curves to determine the range of operating points. The pipe friction loss is given by the Hazen-Williams equation.

$$h_f = \frac{10.44 L Q_{\text{gpm}}^{1.85}}{C^{1.85} d_{\text{in}}^{4.87}}$$

For design, use a roughness coefficient of 140. The discharge pipe friction loss is

$$h_{f(d)} = \frac{(10.44)(2000 \text{ ft}) Q_{\text{gpm}}^{1.85}}{(140)^{1.85}(12 \text{ in})^{4.87}}$$
$$= 12.4 \times 10^{-6} Q_{\text{gpm}}^{1.85} \text{ ft}$$

The suction pipe friction loss is

$$h_{f(s)} = \frac{(10.44)(1500 \text{ ft}) Q_{\text{gpm}}^{1.85}}{(140)^{1.85}(16 \text{ in})^{4.87}}$$
$$= 2.3 \times 10^{-6} Q_{\text{gpm}}^{1.85} \text{ ft}$$

The total friction loss is

$$h_{f(d)} + h_{f(s)} = 12.4 \times 10^{-6} Q_{\text{gpm}}^{1.85}$$
$$+ 2.3 \times 10^{-6} Q_{\text{gpm}}^{1.85}$$
$$= 14.7 \times 10^{-6} Q_{\text{gpm}}^{1.85}$$

At the various flows, the discharge and suction friction losses are

flow rate (gpm)	pump total dynamic head (ft)	total friction loss (ft)
500	96	1.4
1000	88	5.2
1500	76	11.0
2000	60	18.8
2500	36	28.4

The total dynamic head is

$$h_t = h_{z(d)} - h_{z(s)} + h_{f(d)} + h_{f(s)}$$

The discharge static head is

$$h_{z(d)} = 160 \text{ ft} - 90 \text{ ft}$$
$$= 70 \text{ ft}$$

The suction static head at low level is

$$h_{z(s)} = 100 \text{ ft} - 90 \text{ ft} = 10 \text{ ft}$$

The total dynamic head at 500 gpm is

$$h_t = 70 \text{ ft} - 10 \text{ ft} + 1.4 \text{ ft}$$
$$= 61.4 \text{ ft} \quad \text{[low level]}$$

The total dynamic heads for other flows are given in the following table.

flow rate (gpm)	pump total dynamic head (ft)	system head at low level (ft)	system head at high level (ft)
500	96	61.4	41.4
1000	88	65.2	45.2
1500	76	71.0	51.0
2000	60	78.8	58.8
2500	36	88.4	68.4

Plot the flow rate against the pump and system heads.

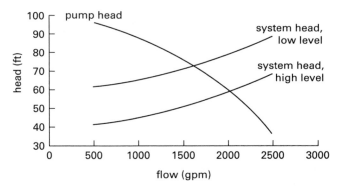

The pump operating points are approximately 1650 gpm at 72 ft and 2000 gpm at 60 ft.

The answer is (D).

220. Statement II is false. This condition would only be true if the flows in pipes 3 and 7 were zero. Statement III is false. The pressure head at node 2 will be the same as the water surface elevation of 100 ft. Statement I is true. Statement IV is true, because the algebraic sum of the energy losses around any loop must be zero.

The answer is (C).

221. The pipe protrudes into the detention pond by two pipe diameters, which approximates a reentrant tube with a coefficient of discharge, C_d, of 0.72. The discharge from the pipe is

$$\dot{V} = C_d A_o \sqrt{2gh}$$

The flow over the spillway is

$$Q = \tfrac{2}{3} C_1 b \sqrt{2g} H^{3/2}$$

Set the flow from the pipe equal to the flow from the spillway, and solve for the head.

$$H = h - 2.5 \text{ ft}$$

$$C_d A_o \sqrt{2gh} = \tfrac{2}{3} C_1 b \sqrt{2g}(h - 2.5 \text{ ft})^{3/2}$$

$$(0.72)\left(\frac{\pi(1 \text{ ft})^2}{4}\right)\sqrt{2gh} = \left(\frac{2}{3}\right)(0.62)(4 \text{ ft})\sqrt{2g}$$

$$\times (h - 2.5 \text{ ft})^{3/2}$$

$$\frac{\sqrt{h}}{(h - 2.5 \text{ ft})^{3/2}} = 2.924$$

Solve by iteration or a calculator solver function.

$$h = 3.22 \text{ ft} \quad (3.2 \text{ ft})$$

The answer is (C).

222. The maximum daily demand for residential/ commercial use is

$$Q_{\text{res}} = UPM$$

U is the per capita daily demand, P is the population, and M is the demand multiplier.

$$Q_{\text{res}} = \frac{\left(180 \; \frac{\text{gal}}{\text{capita-day}}\right)(9000 \text{ capita})(1.8)}{1440 \; \frac{\text{min}}{\text{day}}}$$

$$= 2025 \text{ gpm}$$

The fire demand for essential service structures is

$$C = 18F\sqrt{A_{\text{ft}^2}}$$

The needed fire flow, NFF, is

$$\text{NFF} = CO(1 + X + P)$$

Since the ISO occupancy combustible factor, O, exposure factor, X, and communication factor, P, are 1.0, 0, and 0, respectively, for all zones,

$$\text{NFF} = C(1.0)(1 + 0 + 0) = C$$

Note that the fire flow formula is empirical and not dimensionally consistent. For zone 1, F is 0.8 for class 4. The zone 1 fire demand is

$$\text{NFF} = C = (18)(0.8)\sqrt{20{,}000 \text{ ft}^2} = 2036 \text{ gpm}$$

Rounding to the nearest 250 gpm, the fire demand is 2000 gpm. Check that the fire flow is between 500 gpm and 6000 gpm for class 3, 4, 5, and 6 structures. It is, so use 2000 gpm.

For zone 2, F is 0.6 for class 5. The zone 2 fire demand is

$$\text{NFF} = C = (18)(0.6)\sqrt{50{,}500 \text{ ft}^2} = 2427 \text{ gpm}$$

Rounding to the nearest 250 gpm, the fire demand is 2500 gpm. Check that the fire flow is between 500 gpm and 6000 gpm for class 3, 4, 5, and 6 structures. It is, so use 2500 gpm.

For zone 3, F is 0.8 for class 3. The zone 3 fire demand is

$$\text{NFF} = C = (18)(0.8)\sqrt{40{,}800 \text{ ft}^2} = 2908 \text{ gpm}$$

Rounding to the nearest 250 gpm, the fire demand is 3000 gpm. Check that the fire flow is between 500 and 6000 gpm for class 3, 4, 5, and 6 structures. It is, so use 3000 gpm.

The fire demand based on structure separations for zone 1 (25 ft separation) is 1000 gpm. The fire demand based on structure separations for zone 2 (40 ft separation) is 750 gpm. The fire demand based on structure separations for zone 3 (120 ft separation) is 500 gpm. The controlling fire demand will be the demand for essential service structures. The maximum daily demands are given in the following table.

demand type	maximum daily demand (gpm)		
	zone 1	zone 2	zone 3
residential/ commercial	2025	2025	2025
industrial	80	80	80
fire	2000	2500	3000
maximum	4105	4605	5105

The maximum demand for all uses is zone 3 with 5105 gpm. The required delivery capability that best meets this demand is 5100 gpm.

The answer is (D).

223. The Manning coefficient for a concrete channel is 0.013. The area of the channel section is

$$A = \frac{d^2}{\tan 45°}$$
$$= d^2$$

The hydraulic radius is

$$R = \frac{d \cos 45°}{2}$$
$$= 0.353d$$

From the Manning equation, the normal depth is

$$Q = \left(\frac{1.49}{n}\right) A R^{2/3} \sqrt{S_0}$$
$$400 \frac{\text{ft}^3}{\text{sec}} = \left(\frac{1.49}{0.013}\right) d^2 (0.353d)^{2/3} \sqrt{0.003}$$
$$d = 6.17 \text{ ft}$$

Since the normal depth does not equal the actual depth, the flow is not uniform. The velocity at the control structure is

$$v_1 = \frac{Q}{A_1} = \frac{400 \frac{\text{ft}^3}{\text{sec}}}{(5.2 \text{ ft})^2} = 14.79 \text{ ft/sec}$$

The energy head at the control structure is

$$E_1 = D_1 + \frac{v_1^2}{2g} = 5.2 \text{ ft} + \frac{\left(14.79 \frac{\text{ft}}{\text{sec}}\right)^2}{(2)\left(32.2 \frac{\text{ft}}{\text{sec}^2}\right)}$$
$$= 8.60 \text{ ft}$$

The hydraulic radius at the control structure is

$$R = \frac{d \cos 45°}{2} = \frac{(5.2 \text{ ft})\cos 45°}{2} = 1.84 \text{ ft}$$

Compute the energy head for each 0.2 ft water depth increment. The velocity for a water depth of $D_2 = 5.4$ ft is

$$v_2 = \frac{Q}{A_2} = \frac{400 \frac{\text{ft}^3}{\text{sec}}}{(5.4 \text{ ft})^2} = 13.72 \text{ ft/sec}$$

The energy head at D_2 is

$$E_2 = D_2 + \frac{v_2^2}{2g} = 5.4 \text{ ft} + \frac{\left(13.72 \frac{\text{ft}}{\text{sec}}\right)^2}{(2)\left(32.2 \frac{\text{ft}}{\text{sec}^2}\right)}$$
$$= 8.32 \text{ ft}$$

The hydraulic radius at D_2 is

$$R_2 = \frac{d \cos 45°}{2} = \frac{(5.4 \text{ ft})\cos 45°}{2}$$
$$= 1.91 \text{ ft}$$

The average velocity between D_1 and D_2 is

$$v_{\text{ave}} = \frac{v_1 + v_2}{2} = \frac{14.79 \frac{\text{ft}}{\text{sec}} + 13.72 \frac{\text{ft}}{\text{sec}}}{2} = 14.26 \text{ ft/sec}$$

The average hydraulic radius between D_1 and D_2 is

$$R_{\text{ave}} = \frac{R_1 + R_2}{2} = \frac{1.84 \text{ ft} + 1.91 \text{ ft}}{2} = 1.88 \text{ ft}$$

The average slope of the energy grade line is

$$S_{\text{ave}} = \left(\frac{n v_{\text{ave}}}{1.49 R_{\text{ave}}^{2/3}}\right)^2 = \left(\frac{(0.013)\left(14.26 \frac{\text{ft}}{\text{sec}}\right)}{(1.49)(1.88 \text{ ft})^{2/3}}\right)^2$$
$$= 0.00667$$

The length between D_1 and D_2 is

$$L = \frac{E_1 - E_2}{S_1 - S_0} = \frac{8.60 \text{ ft} - 8.32 \text{ ft}}{0.00667 - 0.003} = 76.3 \text{ ft}$$

The remaining computations are given in *Table for Solution 223.*

The depth at the highway crossing, at a distance of 345 ft upstream from the control structure, is 6.0 ft.

The answer is (C).

Table for Solution 223

depth, D (ft)	velocity, v (ft/sec)	head, $v^2/2g$ (ft)	energy, E (ft)	hydraulic radius, R (ft)	average velocity, v_{ave} (ft/sec)	average hydraulic radius, R_{ave} (ft)	average slope, S_{ave} (ft/ft)	change in energy, ΔE (ft)	length, L (ft)	cumulative length, L_{cum} (ft)
5.2	14.79	3.40	8.60	1.84	–	–	–	–	–	0
–	–	–	–	–	14.26	1.88	0.00667	0.28	76.3	–
5.4	13.72	2.92	8.32	1.91	–	–	–	–	–	76
–	–	–	–	–	13.24	1.95	0.00545	0.19	78	–
5.6	12.76	2.53	8.13	1.98	–	–	–	–	–	154
–	–	–	–	–	12.33	2.02	0.00451	0.13	86	–
5.8	11.89	2.20	8.00	2.05	–	–	–	–	–	240
–	–	–	–	–	11.50	2.09	0.00375	0.08	106	–
6.0	11.11	1.92	7.92	2.12	–	–	–	–	–	346

224. The net positive suction head available is

$$\text{NPSHA} = h_{atm} + h_{z(s)} - h_{f(s)} - h_{vp}$$

The atmospheric pressure at 5000 ft altitude is 12 psi. The atmospheric head is

$$h_{atm} = \frac{p}{\gamma} = \frac{\left(12 \frac{lbf}{in^2}\right)\left(12 \frac{in}{ft}\right)^2}{62.4 \frac{lbf}{ft^3}}$$

$$= 27.7 \text{ ft}$$

The vapor pressure of water at 90°F is 1.62 ft. The friction loss in the suction piping is 3.6 ft. The NPSHA is

$$\text{NPSHA} = 1.3\,\text{NPSHR} = (1.3)(10.3 \text{ ft}) = 13.39 \text{ ft}$$

The elevation difference is

$$h_{z(s)} = \text{NPSHA} - h_{atm} + h_{f(s)} + h_{vp}$$
$$= 13.39 \text{ ft} - 27.7 \text{ ft} + 3.6 \text{ ft} + 1.62 \text{ ft}$$
$$= -9.09 \text{ ft} \quad (-9.0 \text{ ft})$$

The pump centerline must be no more than 9 ft above the water level in the pit.

The answer is (B).

225. The area of the rectangular channel is

$$A = dw = d_1 w = d_1(9 \text{ ft})$$

The hydraulic radius is

$$R = \frac{A}{P} = \frac{dw}{2d + w} = \frac{d_1(9 \text{ ft})}{2d_1 + 9 \text{ ft}}$$

The normal depth is

$$Q = \left(\frac{1.49}{n}\right) A R^{2/3} \sqrt{S}$$

$$530 \frac{ft^3}{sec} = \left(\frac{1.49}{0.01}\right)(9 \text{ ft})d_1 \left(\frac{d_1(9 \text{ ft})}{2d_1 + 9 \text{ ft}}\right)^{2/3} \sqrt{0.004}$$

$$d_1 = 3.84 \text{ ft}$$

The upstream velocity is

$$v_1 = \frac{Q}{A} = \frac{530 \frac{ft^3}{sec}}{(9 \text{ ft})(3.84 \text{ ft})} = 15.34 \text{ ft/sec}$$

The Froude number for a rectangular channel is

$$\text{Fr} = \frac{v}{\sqrt{gd_1}} = \frac{15.34 \frac{ft}{sec}}{\sqrt{\left(32.2 \frac{ft}{sec^2}\right)(3.84 \text{ ft})}} = 1.379$$

The Froude number is greater than 1; therefore, the flow is supercritical.

The downstream depth is

$$\frac{d_2}{d_1} = \tfrac{1}{2}\left(\sqrt{1 + 8(\text{Fr})^2} - 1\right)$$

$$d_2 = \left(\frac{3.84 \text{ ft}}{2}\right)\left(\sqrt{1 + (8)(1.379)^2} - 1\right) = 5.81 \text{ ft}$$

The downstream flow is subcritical.

The energy loss is

$$\Delta E = \frac{(d_2 - d_1)^3}{4d_1 d_2}$$

$$= \frac{(5.81 \text{ ft} - 3.84 \text{ ft})^3}{(4)(3.84 \text{ ft})(5.81 \text{ ft})}$$

$$= 0.086 \text{ ft} \quad (0.09 \text{ ft})$$

The answer is (A).

226. The average discharge from the tank is

$$\dot{V}_{ave} = \frac{V}{t} = \frac{6.66 \text{ ft}^3}{2.8 \text{ sec}}$$
$$= 2.38 \text{ ft}^3/\text{sec}$$

The peak discharge from the tank is

$$\dot{V}_{peak} = \left(2.38 \; \frac{\text{ft}^3}{\text{sec}}\right)\left(\frac{150\%}{100\%}\right)$$
$$= 3.57 \text{ ft}^3/\text{sec}$$

The area of the throat is

$$A = \frac{\pi D^2}{4} = \frac{\pi(6 \text{ in})^2}{(4)\left(12 \; \frac{\text{in}}{\text{ft}}\right)^2}$$
$$= 0.196 \text{ ft}^2$$

The density of mercury is 848.6 lbm/ft^3. The density of water at 50°F is 62.41 lbm/ft^3. The flow coefficient is

$$C_f = \frac{\dot{V}_{peak}}{A\sqrt{\dfrac{2g(\rho_m - \rho_w)h}{\rho_w}}}$$

$$= \frac{3.57 \; \dfrac{\text{ft}^3}{\text{sec}}}{(0.196 \text{ ft}^2)\sqrt{\dfrac{\begin{array}{c}(2)\left(32.2 \; \dfrac{\text{ft}}{\text{sec}^2}\right) \\ \times \left(848 \; \dfrac{\text{lbm}}{\text{ft}^3} - 62.41 \; \dfrac{\text{lbm}}{\text{ft}^3}\right) \\ \times (8 \text{ in}) \end{array}}{\left(62.41 \; \dfrac{\text{lbm}}{\text{ft}^2}\right)\left(12 \; \dfrac{\text{in}}{\text{ft}}\right)}}}$$

$$= 0.783 \quad (0.78)$$

The answer is (A).

227. From a diagram of hydraulic elements of circular sections for a Manning roughness constant of $n = 0.012$, the optimum discharge occurs with the depth of flow-to-diameter ratio, d/D, of 0.93. At this ratio, the discharge ratio is

$$\frac{Q}{Q_f} = 1.075$$

The optimum depth is 0.93 times the diameter of the pipe. When critical velocity occurs at optimum depth, the discharge is a maximum for the available energy. Optimum discharge is, therefore, obtained by making the critical depth the same as the optimum depth. For a 24 in pipe, the critical depth is

$$d_c = 0.93D$$
$$= \frac{(0.93)(24 \text{ in})}{12 \; \dfrac{\text{in}}{\text{ft}}}$$
$$= 1.86 \text{ ft}$$

From the hydraulic elements diagram for $d/D = 0.93$,

$$\frac{A}{A_f} = 0.96$$
$$\frac{R}{R_f} = 1.17$$

The area is

$$A = \frac{(0.96)\left(\dfrac{\pi(24 \text{ in})^2}{4}\right)}{\left(12 \; \dfrac{\text{in}}{\text{ft}}\right)^2}$$
$$= 3.02 \text{ ft}^2$$

The full hydraulic radius is

$$R_{full} = \frac{A_{full}}{P_{full}} = \frac{\pi D^2}{4\pi D} = \frac{D}{4}$$
$$= \frac{2 \text{ ft}}{4}$$
$$= 0.5 \text{ ft}$$

The hydraulic radius at critical depth is

$$\frac{R}{R_{full}} = 1.17$$
$$R = (0.5 \text{ ft})(1.17)$$
$$= 0.585 \text{ ft}$$

The critical slope can be found from the Manning equation.

$$S = \left(\frac{Qn}{1.49AR^{2/3}}\right)^2$$
$$= \left(\frac{\left(30 \; \dfrac{\text{ft}^3}{\text{sec}}\right)(0.012)}{(1.49)(3.02 \text{ ft}^2)(0.585 \text{ ft})^{2/3}}\right)^2$$
$$= 0.013 \text{ ft/ft}$$

The answer is (B).

228. Since depth and velocity are inversely proportional, the velocity can be scaled down from the point of minimum depth.

$$v_o d_o = v_1 d_1$$

$$\left(62 \ \frac{\text{ft}}{\text{sec}}\right)(4.08 \ \text{ft}) = v_1 (4.57 \ \text{ft})$$

$$v_1 = 55.35 \ \text{ft/sec}$$

The depths immediately before and after a hydraulic jump are the conjugate depths, d_1 and d_2. One can be calculated from the other.

$$d_2 = -\frac{d_1}{2} + \sqrt{\frac{2v_1^2 d_1}{g} + \frac{d_1^2}{4}}$$

$$= -\frac{4.57 \ \text{ft}}{2}$$

$$+ \sqrt{\frac{(2)\left(55.35 \ \dfrac{\text{ft}}{\text{sec}}\right)^2 (4.57 \ \text{ft})}{32.2 \ \dfrac{\text{ft}}{\text{sec}^2}} + \frac{(4.57 \ \text{ft})^2}{4}}$$

$$= 27.29 \ \text{ft} \quad (27 \ \text{ft})$$

The answer is (D).

229. The normal annual precipitation of station D relative to stations A, B, and C is

$$\frac{\text{station D}}{\text{station A}} = \frac{50.3 \ \text{in}}{40.5 \ \text{in}}$$
$$= 1.24$$
$$\frac{\text{station D}}{\text{station B}} = \frac{50.3 \ \text{in}}{32.8 \ \text{in}}$$
$$= 1.53$$
$$\frac{\text{station D}}{\text{station C}} = \frac{50.3 \ \text{in}}{60.7 \ \text{in}}$$
$$= 0.829$$

The normal annual precipitation of station D, which is missing a value for Storm Alpha precipitation, varies by more than 10% from the other stations. The normal-ratio method, which applies if the precipitation difference between locations is more than 10%, can be used to estimate the unknown precipitation at station D. The missing Storm Alpha precipitation is

$$P_x = \frac{1}{3}\left(\left(\frac{N_x}{N_A}\right)P_A + \left(\frac{N_x}{N_B}\right)P_B + \left(\frac{N_x}{N_C}\right)P_C\right)$$

$$= \left(\frac{1}{3}\right)\left(\begin{array}{l}\left(\dfrac{50.3 \ \text{in}}{40.5 \ \text{in}}\right)(3.20 \ \text{in}) \\[2mm] + \left(\dfrac{50.3 \ \text{in}}{32.8 \ \text{in}}\right)(2.70 \ \text{in}) \\[2mm] + \left(\dfrac{50.3 \ \text{in}}{60.7 \ \text{in}}\right)(3.90 \ \text{in}) \end{array}\right)$$

$$= 3.78 \ \text{in}$$

The average precipitation over the watershed is

$$P_{\text{avg}} = \frac{P_A A_A + P_B A_B + P_C A_C + P_D A_D}{A_A + A_B + A_C + A_D}$$

$$= \dfrac{\begin{array}{l}(3.20 \ \text{in})(2000 \ \text{ac}) \\ + (2.70 \ \text{in})(2500 \ \text{ac}) \\ + (3.90 \ \text{in})(2800 \ \text{ac}) \\ + (3.78 \ \text{in})(2400 \ \text{ac})\end{array}}{2000 \ \text{ac} + 2500 \ \text{ac} + 2800 \ \text{ac} + 2400 \ \text{ac}}$$

$$= 3.41 \ \text{in}$$

The rainfall intensity is

$$I = \frac{P_{\text{avg}}}{t} = \left(\frac{3.41 \ \text{in}}{57 \ \text{min}}\right)\left(60 \ \frac{\text{min}}{\text{hr}}\right)$$
$$= 3.59 \ \text{in/hr}$$

From the given intensity-duration-frequency curve, the return frequency is most nearly 50 yr.

The answer is (D).

230. The Doorenbos and Pruitt vapor transport coefficient is

$$B = 0.0027\left(1 + \frac{u}{100}\right)$$

u is the 24 hr wind run in km/d.

$$B = (0.0027)\left(1 + \frac{260 \ \dfrac{\text{km}}{\text{d}}}{100}\right)$$

$$= 0.009\,72 \ \text{mm/d·Pa}$$

The saturated vapor pressure of water vapor over liquid water at 20°C, e_{as}, is 2.338 kPa, or 2338 Pa. The relative humidity, R_h, is 30%, or 0.30.

$$e_a = R_h e_{\text{as}} = (0.30)(2338 \ \text{Pa})$$
$$= 701.4 \ \text{Pa}$$

The evaporation rate from aerodynamic vapor transport is

$$E_a = B(e_{as} - e_a)$$
$$= \left(0.009\,72 \ \frac{\text{mm}}{\text{Pa·d}}\right)(2338 \ \text{Pa} - 701.4 \ \text{Pa})$$
$$= 15.91 \ \text{mm/d}$$

The latent heat of vaporization is

$$l_v = 2500 - 2.36 \, T$$
$$= 2500 - (2.36)(20°\text{C})$$
$$= 2452.8 \ \text{kJ/kg}$$

The density of water at $20°C$ is 998.23 kg/m^3. The evaporation rate from radiation is

$$E_r = \frac{R_n}{l_v \rho_w}$$

$$= \frac{\left(250 \, \frac{\text{W}}{\text{m}^2}\right)\left(10^3 \, \frac{\text{mm}}{\text{m}}\right)\left(86\,400 \, \frac{\text{s}}{\text{d}}\right)}{\left(2452.8 \, \frac{\text{kJ}}{\text{kg}}\right)\left(1000 \, \frac{\text{J}}{\text{kJ}}\right)\left(998.23 \, \frac{\text{kg}}{\text{m}^3}\right)}$$

$$= 8.82 \text{ mm/d}$$

The evapotranspiration for the reference crop is the sum of the radiation evaporation rate, multiplied by its weighting factor, and the vapor transport evaporation rate, multiplied by its weighting factor.

$$E_{\text{tr}} = \frac{\Delta}{\Delta + \gamma} E_r + \frac{\gamma}{\Delta + \gamma} E_a$$

$$= (0.7)\left(8.82 \, \frac{\text{mm}}{\text{d}}\right) + (0.3)\left(15.91 \, \frac{\text{mm}}{\text{d}}\right)$$

$$= 10.95 \text{ mm/d}$$

The evapotranspiration for the alfalfa is

$$E_t = k_s k_c E_{\text{tr}}$$

$$= (0.80)(0.35)\left(10.95 \, \frac{\text{mm}}{\text{d}}\right)$$

$$= 3.07 \text{ mm/d} \quad (3 \text{ mm/d})$$

The answer is (B).

231. Determine the intensity of precipitation for each subduration and arrange them in descending order. The intensity for the 1.9 cm rainfall depth for 5 min duration is

$$I = \left(\frac{1.9 \text{ cm}}{5 \text{ min}}\right)\left(60 \, \frac{\text{min}}{\text{h}}\right)$$

$$= 22.8 \text{ cm/h}$$

The probability is

$$P = \frac{M}{n + 1 \text{ yr}}$$

M is the order number, and n is the number of years.

$$P = \frac{M}{10 \text{ yr} + 1 \text{ yr}}$$

$$= \frac{M}{11 \text{ yr}}$$

The calculations for the 5 min and 20 min durations are given in the following table.

5 min duration

depth (cm)	intensity (cm/h)	order number, M	probability
1.9	22.8	1	0.09
1.8	21.6	2	0.18
1.5	18.0	3	0.27
1.3	15.6	4	0.36
1.2	14.4	5	0.45
1.2	14.4	6	0.55
1.0	12.0	7	0.64
0.9	10.8	8	0.73
0.8	9.6	9	0.82
0.7	8.4	10	0.91

20 min duration

depth (cm)	intensity (cm/h)	order number, M	probability
3.2	9.6	1	0.09
3.1	9.3	2	0.18
2.9	8.7	3	0.27
2.8	8.4	4	0.36
2.6	7.8	5	0.45
2.5	7.5	6	0.55
2.4	7.2	7	0.64

5 min duration

depth (cm)	intensity (cm/h)	order number, M	probability
2.4	7.2	8	0.73
2.3	6.9	9	0.82
2.1	6.3	10	0.91

Determine the probability of the 2-yr and 10-yr return period.

$$P = \frac{1}{T}$$

$$P_{\text{2-yr}} = \frac{1}{2 \text{ yr}} = 0.5$$

$$P_{\text{10-yr}} = \frac{1}{10 \text{ yr}} = 0.1$$

The intensity for the 2-yr ($P_{\text{2-yr}} = 0.5$) return period for the 5 min subduration is found by interpolation in the table.

$$x = 14.4 \, \frac{\text{cm}}{\text{h}} - \left(14.4 \, \frac{\text{cm}}{\text{h}} - 14.4 \, \frac{\text{cm}}{\text{h}}\right)\left(\frac{0.5 - 0.45}{0.55 - 0.45}\right)$$

$$= 14.4 \text{ cm/h}$$

The intensity for the 10-yr ($P_{\text{10-yr}} = 0.1$) return period for the 5 min subduration is found by interpolation.

$$x = 22.8 \, \frac{\text{cm}}{\text{h}} - \left(22.8 \, \frac{\text{cm}}{\text{h}} - 21.6 \, \frac{\text{cm}}{\text{h}}\right)\left(\frac{0.1 - 0.09}{0.18 - 0.09}\right)$$

$$= 22.67 \text{ cm/h}$$

The intensity for the 2-yr ($P_{2\text{-yr}} = 0.5$) return period for the 20 min subduration is found by interpolation.

$$x = 7.8 \ \frac{\text{cm}}{\text{h}} - \left(7.8 \ \frac{\text{cm}}{\text{h}} - 7.5 \ \frac{\text{cm}}{\text{h}}\right)\left(\frac{0.5 - 0.45}{0.55 - 0.45}\right)$$
$$= 7.65 \ \text{cm/h}$$

The intensity for the 10-yr ($P_{10\text{-yr}} = 0.1$) return period for the 20 min subduration is found by interpolation.

$$x = 9.6 \ \frac{\text{cm}}{\text{h}} - \left(9.6 \ \frac{\text{cm}}{\text{h}} - 9.3 \ \frac{\text{cm}}{\text{h}}\right)\left(\frac{0.1 - 0.09}{0.18 - 0.09}\right)$$
$$= 9.57 \ \text{cm/h}$$

The intensities for the required return periods for each subduration are given in the following table.

subduration	2-yr return intensity (cm/h)	10-yr return intensity (cm/h)
5 min	14.4	22.67
20 min	7.65	9.57

For a 15 min storm, the 2-yr intensity for the 2-yr period is

$$x = 14.4 \ \frac{\text{cm}}{\text{h}} - \left(14.4 \ \frac{\text{cm}}{\text{h}} - 7.65 \ \frac{\text{cm}}{\text{h}}\right)$$
$$\times \left(\frac{15 \ \text{min} - 5 \ \text{min}}{20 \ \text{min} - 5 \ \text{min}}\right)$$
$$= 9.90 \ \text{cm/h}$$

For the 10-yr return period,

$$x = 22.67 \ \frac{\text{cm}}{\text{h}} - \left(22.67 \ \frac{\text{cm}}{\text{h}} - 9.57 \ \frac{\text{cm}}{\text{h}}\right)$$
$$\times \left(\frac{15 \ \text{min} - 5 \ \text{min}}{20 \ \text{min} - 5 \ \text{min}}\right)$$
$$= 13.94 \ \text{cm/h}$$

The percent increase is

$$\left(\frac{13.94 \ \frac{\text{cm}}{\text{h}} - 9.90 \ \frac{\text{cm}}{\text{h}}}{9.90 \ \frac{\text{cm}}{\text{h}}}\right) \times 100\% = 40.8\% \quad (40\%)$$

The answer is (D).

232. Sheet flow depth and unit width are shown in the following illustration.

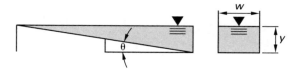

The intensity is

$$i = \frac{1.3 \ \frac{\text{in}}{\text{hr}}}{\left(12 \ \frac{\text{in}}{\text{ft}}\right)\left(3600 \ \frac{\text{sec}}{\text{hr}}\right)}$$
$$= 30 \times 10^{-6} \ \text{ft/sec}$$

The slope angle θ is

$$\theta = \tan^{-1} S = \tan^{-1} 0.06$$
$$= 3.43°$$

The kinematic viscosity at 60°F is $1.217 \times 10^{-5} \ \text{ft}^2/\text{sec}$. The discharge per unit width is

$$q_o = vy$$
$$= (i - f)L_o \cos \theta$$
$$= \left(30 \times 10^{-6} \ \frac{\text{ft}}{\text{sec}} - 0\right)(150 \ \text{ft})\cos 3.43°$$
$$= 4.5 \times 10^{-3} \ \text{ft}^3/\text{sec-ft}$$

For sheet flow with unit width, the hydraulic radius is

$$R = \frac{A}{P} = \frac{wy}{w} = y$$

The Reynolds number is

$$\text{Re} = \frac{D_e v}{\nu} = \frac{4Rv}{\nu}$$
$$= \frac{4yv}{\nu}$$
$$= \frac{4q_o}{\nu}$$
$$= \frac{(4)\left(4.5 \times 10^{-3} \ \frac{\text{ft}^3}{\text{sec-ft}}\right)}{1.217 \times 10^{-5} \ \frac{\text{ft}^2}{\text{sec}}}$$
$$= 1479$$

The Reynolds number is less than 2000, so the flow is laminar. For laminar flow under rainfall, the friction factor increases with rainfall intensity according to the following equations.

$$C_L = 96 + 108i^{0.4}$$
$$= 96 + (108)\left(1.3 \ \frac{\text{in}}{\text{hr}}\right)^{0.4}$$
$$= 216$$

The friction factor is

$$f = \frac{C_L}{\text{Re}} = \frac{216}{1479}$$
$$= 0.146$$

The depth of sheet flow is

$$y = \left(\frac{fq_o^2}{8gS}\right)^{1/3}$$

$$= \left(\frac{(0.146)\left(4.5 \times 10^{-3} \; \frac{\text{ft}^3}{\text{sec-ft}}\right)^2}{(8)\left(32.2 \; \frac{\text{ft}}{\text{sec}^2}\right)(0.06)}\right)^{1/3}$$

$$= 0.00576 \; \text{ft}$$

The velocity is

$$v = \frac{q_o}{y} = \frac{4.5 \times 10^{-3} \; \frac{\text{ft}^3}{\text{sec-ft}}}{0.00576 \; \text{ft}}$$

$$= 0.781 \; \text{ft/sec} \quad (0.80 \; \text{ft/sec})$$

The answer is (B).

233. At steady flow, the discharge can be determined by summing the flow in each section of the channel and flood plain. Consider the sections as parallel flows with no wetted perimeter between the interface.

For the channel, the area is

$$A_{\text{channel}} = d_{\text{channel}} w_{\text{channel}}$$

$$= (8 \; \text{ft})(50 \; \text{ft})$$

$$= 400 \; \text{ft}^2$$

The wetted perimeter is

$$P_{\text{channel}} = (d_{\text{channel}} - d_{\text{west}}) + w_{\text{channel}}$$

$$+ (d_{\text{channel}} - d_{\text{east}})$$

$$= (8 \; \text{ft} - 2 \; \text{ft}) + 50 \; \text{ft} + (8 \; \text{ft} - 1 \; \text{ft})$$

$$= 63 \; \text{ft}$$

The hydraulic radius is

$$R_{\text{channel}} = \frac{A_{\text{channel}}}{P_{\text{channel}}} = \frac{400 \; \text{ft}^2}{63 \; \text{ft}}$$

$$= 6.35 \; \text{ft}$$

The discharge is

$$Q_{\text{channel}} = \left(\frac{1.49}{n}\right) A_{\text{channel}} R_{\text{channel}}^{2/3} \sqrt{S_0}$$

$$= \left(\frac{1.49}{0.025}\right)(400 \; \text{ft}^2)(6.35 \; \text{ft})^{2/3}\sqrt{0.002}$$

$$= 3656 \; \text{ft}^3/\text{sec}$$

For the west flood plain, the area is

$$A_{\text{west}} = (2 \; \text{ft})(300 \; \text{ft})$$

$$= 600 \; \text{ft}^2$$

The wetted perimeter is

$$P_{\text{west}} = d_{\text{west}} + w_{\text{west}}$$

$$= 2 \; \text{ft} + 300 \; \text{ft}$$

$$= 302 \; \text{ft}$$

The hydraulic radius is

$$R_{\text{west}} = \frac{600 \; \text{ft}^2}{302 \; \text{ft}}$$

$$= 1.99 \; \text{ft}$$

The discharge is

$$Q_{\text{west}} = \left(\frac{1.49}{0.035}\right)(600 \; \text{ft}^2)(1.99 \; \text{ft})^{2/3}\sqrt{0.002}$$

$$= 1807.3 \; \text{ft}^3/\text{sec}$$

For the east flood plain, the area is

$$A_{\text{east}} = (1 \; \text{ft})(400 \; \text{ft})$$

$$= 400 \; \text{ft}^2$$

The wetted perimeter is

$$P_{\text{east}} = d_{\text{east}} + w_{\text{east}} = 1 \; \text{ft} + 400 \; \text{ft}$$

$$= 401 \; \text{ft}$$

The hydraulic radius is

$$R_{\text{east}} = \frac{400 \; \text{ft}^2}{401 \; \text{ft}}$$

$$= 0.99 \; \text{ft}$$

The discharge is

$$Q_{\text{east}} = \left(\frac{1.49}{0.060}\right)(400 \; \text{ft}^2)(0.99 \; \text{ft})^{2/3}\sqrt{0.002}$$

$$= 441.3 \; \text{ft}^3/\text{sec}$$

The total flood flow is

$$Q_{\text{total}} = Q_{\text{channel}} + Q_{\text{west}} + Q_{\text{east}}$$

$$= 3656 \; \frac{\text{ft}^3}{\text{sec}} + 1807.3 \; \frac{\text{ft}^3}{\text{sec}} + 441.3 \; \frac{\text{ft}^3}{\text{sec}}$$

$$= 5905 \; \text{ft}^3/\text{sec} \quad (6000 \; \text{ft}^3/\text{sec})$$

The answer is (C).

234. For the first subduration, the demand consumption is

$$D = (\text{design demand rate})(\text{subduration})$$
$$= \left(995\,000\ \frac{\text{m}^3}{\text{d}}\right)(7\ \text{d})$$
$$= 6\,965\,000\ \text{m}^3$$

The evaporation/seepage loss is

$$L = (\text{evaporation/seepage rate})(\text{reservoir area})$$
$$= \frac{(30\ \text{mm})(4\ \text{km}^2)\left(1000\ \frac{\text{m}}{\text{km}}\right)^2}{1000\ \frac{\text{mm}}{\text{m}}}$$
$$= 120\,000\ \text{m}^3$$

The inflow volume is

$$I = (\text{average inflow rate})(\text{subduration})$$
$$= \left(120\,000\ \frac{\text{m}^3}{\text{d}}\right)(7\ \text{d})$$
$$= 840\,000\ \text{m}^3$$

subduration (d)	average inflow rate (Mm³/d)	inflow volume (Mm³)	evaporation/ seepage rate (mm)
7	0.120	0.840	30
30	0.150	4.500	120
60	0.180	10.800	250
120	0.240	28.800	500
180	0.560	100.800	780
365	1.800	657.000	1600

subduration (d)	evaporation/ seepage volume (Mm³)	design demand rate (Mm³)	design demand volume (Mm³)	required storage (Mm³)
7	0.120	0.995	6.965	6.245
30	0.480	0.990	29.700	25.680
60	1.000	0.980	58.800	49.000
120	2.000	0.960	115.200	88.400
180	3.120	0.940	169.200	71.520
365	6.400	0.900	328.500	-322.100

The required storage volume can be determined from a mass balance with respect to the reservoir. A volume basis can be used because the density of water is essentially constant for these conditions. The required storage is the demand plus losses minus the inflow. The net storage required for the 7 d subduration is

$$S = D + L - I$$
$$= 6\,965\,000\ \frac{\text{m}^3}{\text{d}} + 120\,000\ \frac{\text{m}^3}{\text{d}} - 840\,000\ \frac{\text{m}^3}{\text{d}}$$
$$= 6\,245\,000\ \text{m}^3$$

Calculations for the other subdurations are given in the table. The largest storage volume of the subdurations analyzed is $88\,400\,000\ \text{m}^3$ ($88\ \text{Mm}^3$) for the 120 d subduration, which will contain the smaller volumes of other subdurations.

The answer is (D).

235. The volume of water treated is

$$V = Qt$$
$$= \left(2\ \frac{\text{m}^3}{\text{s}}\right)\left(86\,400\ \frac{\text{s}}{\text{d}}\right)\left(1000\ \frac{\text{L}}{\text{m}^3}\right)$$
$$= 172.8 \times 10^6\ \text{L/d}$$

The mass of alum sludge is

$$m_{\text{alum}} = QC(\text{efficiency})$$
$$= \frac{\left(172.8 \times 10^6\ \frac{\text{L}}{\text{d}}\right)\left(12\ \frac{\text{mg alum dose}}{\text{L}}\right)}{10^6\ \frac{\text{mg}}{\text{kg}}}$$
$$\times \left(0.46\ \frac{\text{kg alum sludge}}{\text{kg alum dose}}\right)(0.96)$$
$$= 915.7\ \text{kg/d alum sludge}$$

The equivalent concentration of TSS removed is

$$\Delta\text{TSS} = \Delta\text{NTU}^{1.2}$$
$$= (5\ \text{NTU} - 1\ \text{NTU})^{1.2}$$
$$= 5.28\ \text{mg/L}$$

The mass of TSS removed is

$$m_{\text{TSS}} = QC(\text{efficiency})$$
$$= \frac{\left(172.8 \times 10^6\ \frac{\text{L}}{\text{d}}\right)\left(5.28\ \frac{\text{mg}}{\text{L}}\right)(0.96)}{10^6\ \frac{\text{mg}}{\text{kg}}}$$
$$= 875.9\ \text{kg/d}$$

The mass of clay removed is

$$m_{\text{clay}} = QC(\text{efficiency})$$

$$= \frac{\left(172.8 \times 10^6 \; \frac{\text{L}}{\text{d}}\right)\left(3 \; \frac{\text{mg}}{\text{L}}\right)(0.96)}{10^6 \; \frac{\text{mg}}{\text{kg}}}$$

$$= 497.7 \text{ kg/d}$$

The total mass of sludge generated following sedimentation is

$$m_{\text{sludge}} = m_{\text{alum}} + m_{\text{TSS}} + m_{\text{clay}}$$

$$= 915.7 \; \frac{\text{kg}}{\text{d}} + 875.9 \; \frac{\text{kg}}{\text{d}} + 497.7 \; \frac{\text{kg}}{\text{d}}$$

$$= 2289.3 \text{ kg/d} \quad (2300 \text{ kg/d})$$

The answer is (B).

236. Use the settling curves to determine the percent solids removal for the detention time associated with the midpoint between the curves. The 20% removal curve intersects the x-axis at 24 min.

The overflow rate is

$$v^* = \frac{\text{settling depth}}{\text{settling time}}$$

$$= \left(\frac{10 \text{ ft}}{24 \text{ min}}\right)\left(1440 \; \frac{\text{min}}{\text{day}}\right)\left(7.48 \; \frac{\text{gal}}{\text{ft}^3}\right)$$

$$= 4488 \text{ gal/day-ft}^2$$

The detention time in hours is

$$t = \frac{24 \text{ min}}{60 \; \frac{\text{min}}{\text{hr}}}$$

$$= 0.40 \text{ hr}$$

Project a line vertically from 24 min. The point midway between 20% and 30% removal (25%) corresponds to a depth of 6.4 ft. Similarly, the depths corresponding to other removal percentages for 0.4 hr detention are shown in the following table.

solids removal (%)	depth (ft)
25	6.4
35	3.8
45	2.8
55	2.4
65	2.0
75	1.7

The total percent removed for a detention time of 0.4 hr is

$$R_t = \text{percent removal for detention time}$$
$$+ \left(\frac{\text{depth at midpoint}}{\text{total depth}}\right)$$
$$\times (\text{increment of percent removal})$$

$$= 20\% + \left(\frac{6.4 \text{ ft}}{10 \text{ ft}}\right)(30\% - 20\%)$$

$$+ \left(\frac{3.8}{10 \text{ ft}}\right)(40\% - 30\%)$$

$$+ \left(\frac{2.8 \text{ ft}}{10 \text{ ft}}\right)(50\% - 40\%)$$

$$+ \left(\frac{2.4 \text{ ft}}{10 \text{ ft}}\right)(60\% - 50\%)$$

$$+ \left(\frac{2.0 \text{ ft}}{10 \text{ ft}}\right)(70\% - 60\%)$$

$$+ \left(\frac{1.7 \text{ ft}}{10 \text{ ft}}\right)(80\% - 70\%)$$

$$= 39.1\%$$

Calculations for the other removals and detention times are given in the following table.

detention time (hr)	overflow rate (gal/day-ft²)	solids removal (%)
0.40	4488	39.1
0.63	2835	52.5
0.87	2071	63.7
1.07	1683	70.4
1.30	1381	75.6
1.57	1146	79.1

For 75% removal, the required detention time is 1.3 hr and the overflow rate is 1380 gal/day-ft² (scale-up factor is 1.0).

The required area is

$$A = \frac{Q}{v^*}$$

$$= \frac{1.5 \times 10^6 \; \frac{\text{gal}}{\text{day}}}{1380 \; \frac{\text{gal}}{\text{day-ft}^2}}$$

$$= 1087 \text{ ft}^2$$

The diameter is

$$D = \sqrt{\frac{4A}{\pi}}$$

$$= \sqrt{\frac{(4)(1087 \text{ ft}^2)}{\pi}}$$

$$= 37.2 \text{ ft} \quad (37 \text{ ft})$$

The required depth is

$$h = \frac{Qt}{A}$$

$$= \frac{\left(1.5 \times 10^6 \ \frac{\text{gal}}{\text{day}}\right)(1.3 \ \text{hr})}{(1087 \ \text{ft}^2)\left(24 \ \frac{\text{hr}}{\text{day}}\right)\left(7.48 \ \frac{\text{gal}}{\text{ft}^3}\right)}$$

$$= 9.99 \ \text{ft} \quad (10 \ \text{ft})$$

The answer is (B).

237. Since each layer is uniform, the weight fraction of each layer is 1, and all particles in the layer have the same diameter. The head loss can be found from the Rose equation. Sum the head loss from the individual layers to find the total head loss. The kinematic viscosity at 10°C is 1.371×10^{-6} m²/s. The approach velocity is the filtration rate as velocity.

$$\text{v}_a = \frac{4.5 \ \frac{\text{L}}{\text{s·m}^2}}{1000 \ \frac{\text{L}}{\text{m}^3}}$$

$$= 0.0045 \ \text{m/s}$$

For the sand layer, the Reynolds number is

$$\text{Re} = \frac{\phi d \text{v}_a}{\nu}$$

$$= \frac{(0.8)(1.0 \ \text{mm})\left(0.0045 \ \frac{\text{m}}{\text{s}}\right)}{\left(1.371 \times 10^{-6} \ \frac{\text{m}^2}{\text{s}}\right)\left(1000 \ \frac{\text{mm}}{\text{m}}\right)}$$

$$= 2.626$$

Using the equation developed by T.R. Camp to describe drag on spheres. For a Reynolds number greater than 1 but less than 10,000, the drag coefficient is

$$C_D = \frac{24}{\text{Re}} + \frac{3}{\sqrt{\text{Re}}} + 0.34$$

$$= \frac{24}{2.626} + \frac{3}{\sqrt{2.626}} + 0.34$$

$$= 11.33$$

Using the Rose equation for stratified beds with uniform porosity, the head loss in the sand layer is

$$h_L = \left(\frac{1.067 D \text{v}_a^2}{\phi g \xi^4}\right) \sum \frac{C_D x}{d}$$

D is the depth of weight fraction, and x is the weight fraction of particles (1 for uniform layer).

$$h_L = \frac{(1.067)(280 \ \text{mm})\left(1000 \ \frac{\text{mm}}{\text{m}}\right)}{\left(0.8\right)\left(9.81 \ \frac{\text{m}}{\text{s}^2}\right)(0.50)^4(1.0 \ \text{mm})\left(1000 \ \frac{\text{mm}}{\text{m}}\right)}$$

$$= 0.1397 \ \text{m} \quad (0.140 \ \text{m})$$

For the garnet layer, the Reynolds number is

$$\text{Re} = \frac{(0.8)(0.40 \ \text{mm})\left(0.0045 \ \frac{\text{m}}{\text{s}}\right)}{\left(1.371 \times 10^{-6} \ \frac{\text{m}^2}{\text{s}}\right)\left(1000 \ \frac{\text{mm}}{\text{m}}\right)}$$

$$= 1.05$$

For a Reynolds number greater than 1 but less than 10,000, the drag coefficient is

$$C_D = \frac{24}{1.05} + \frac{3}{\sqrt{1.05}} + 0.34$$

$$= 26.12$$

The head loss in the garnet layer is

$$h_L = \frac{(1.067)(100 \ \text{mm})\left(1000 \ \frac{\text{mm}}{\text{m}}\right)}{\left(0.8\right)\left(9.81 \ \frac{\text{m}}{\text{s}^2}\right)(0.55)^4(0.40 \ \text{mm})\left(1000 \ \frac{\text{mm}}{\text{m}}\right)}$$

$$= 0.1965 \ \text{m} \quad (0.20 \ \text{m})$$

The total head loss is

$$\sum h_L = 0.14 \ \text{m} + 0.20 \ \text{m} = 0.34 \ \text{m}$$

The answer is (D).

238. The mass of water to be treated is

$$m_{\text{water}} = V\rho$$

$$= (1000 \ \text{m}^3)\left(1000 \ \frac{\text{kg}}{\text{m}^3}\right)$$

$$= 1\,000\,000 \ \text{kg}$$

The required mass of hypochlorite with a gravimetric availability of G is

$$m_{\text{hypo}} = \frac{m_{\text{water}} C_{\text{hypo}}}{G}$$

$$= \frac{(1\,000\,000 \text{ kg})\left(0.005 \, \dfrac{\text{kg OCl}}{\text{kg solution}}\right)}{0.70}$$

$$= 7143 \text{ kg} \quad (7000 \text{ kg})$$

The answer is (D).

239. The gross velocity of groundwater flow is

$$v_{\text{gross}} = Ki = K\frac{\Delta H}{L}$$

$$= \left(0.1 \, \frac{\text{m}}{\text{d}}\right)\left(\frac{40 \text{ m} - 0 \text{ m}}{2000 \text{ m}}\right)$$

$$= 2 \times 10^{-3} \text{ m/d}$$

The pore velocity of groundwater flow is

$$v_{\text{pore}} = \frac{v_{\text{gross}}}{n}$$

n represents porosity.

$$v_{\text{pore}} = \frac{2 \times 10^{-3} \, \dfrac{\text{m}}{\text{d}}}{0.25}$$

$$= 8 \times 10^{-3} \text{ m/d}$$

The coefficient of retardation, C_r, is the ratio of the velocity of the centroid of the contaminant plume to the groundwater velocity. The velocity of the contaminant plume is

$$v_{\text{plume}} = C_r v_{\text{pore}}$$

Try the coefficient of retardation for more than 640 d travel time. The velocity of the plume is

$$v_{\text{plume}} = C_r v_{\text{pore}}$$

$$= (0.17)\left(8.0 \times 10^{-3} \, \frac{\text{m}}{\text{d}}\right)$$

$$= 1.36 \times 10^{-3} \text{ m/d}$$

The travel time is

$$t = \frac{L}{v_{\text{plume}}}$$

$$= \frac{2000 \text{ m}}{1.36 \times 10^{-3} \, \dfrac{\text{m}}{\text{d}}}$$

$$= 1.47 \times 10^6 \text{ d}$$

The result that shows the coefficient of retardation used in the preceding equation is appropriate, because the travel time is more than 640 d.

$$t = \frac{1.47 \times 10^6 \text{ d}}{365 \, \dfrac{\text{d}}{\text{yr}}}$$

$$= 4027 \text{ yr} \quad (4000 \text{ yr})$$

The answer is (A).

240. The volume of the basin is

$$V_{\text{basin}} = Qt = \left(3.5 \, \frac{\text{ft}^3}{\text{sec}}\right)(50 \text{ sec})$$

$$= 175 \text{ ft}^3$$

The power required is

$$P = \mu G^2 V_{\text{basin}}$$

$$= \left(2.050 \times 10^{-5} \, \frac{\text{lbf-sec}}{\text{ft}^2}\right)\left(850 \, \frac{\text{ft}}{\text{sec-ft}}\right)^2 (175 \text{ ft}^3)$$

$$= 2592 \text{ lbf-ft/sec}$$

Assume turbulent flow. The power required to drive the impeller is

$$P = \frac{N_P n^3 D^5 \rho}{g_c}$$

The impeller diameter is

$$D = \left(\frac{Pg_c}{N_P n^3 \rho}\right)^{1/5}$$

$$= \left(\frac{\left(2592 \, \dfrac{\text{lbf-ft}}{\text{sec}}\right)\left(32.2 \, \dfrac{\text{lbm-ft}}{\text{lbf-sec}^2}\right)}{(5.5)\left(3 \, \dfrac{\text{rev}}{\text{sec}}\right)^3 \left(62.4 \, \dfrac{\text{lbm}}{\text{ft}^3}\right)}\right)^{1/5}$$

$$= 1.55 \text{ ft} \quad (1.6 \text{ ft})$$

Calculate the Reynolds number to check the turbulent flow assumption.

$$\text{Re} = \frac{D^2 n\rho}{g_c \mu}$$

$$= \frac{(1.6 \text{ ft})^2 \left(3 \, \dfrac{\text{rev}}{\text{sec}}\right)\left(62.4 \, \dfrac{\text{lbm}}{\text{ft}^3}\right)}{\left(32.2 \, \dfrac{\text{lbm-ft}}{\text{lbf-sec}^2}\right)\left(2.050 \times 10^{-5} \, \dfrac{\text{lbf-sec}}{\text{ft}^2}\right)}$$

$$= 726{,}000$$

The Reynolds number is greater than 10,000, so the assumption of turbulent flow is appropriate.

The answer is (B).